G. Baldwin (Gerard Baldwin) Brown

From Schola to cCathedral

A Study of Early Christian Architecture and its Relation to the life of the...

G. Baldwin (Gerard Baldwin) Brown

From Schola to cCathedral
A Study of Early Christian Architecture and its Relation to the life of the...

ISBN/EAN: 9783337003258

Printed in Europe, USA, Canada, Australia, Japan

Cover: Foto ©berggeist007 / pixelio.de

More available books at **www.hansebooks.com**

FROM

SCHOLA TO CATHEDRAL

A Study of

EARLY CHRISTIAN ARCHITECTURE

AND ITS RELATION TO THE LIFE OF THE CHURCH

BY

G. BALDWIN BROWN, M.A.

LATE FELLOW OF BRASENOSE COLLEGE, OXFORD, AND WATSON-GORDON
PROFESSOR OF FINE ART IN THE UNIVERSITY OF EDINBURGH

EDINBURGH: DAVID DOUGLAS

1886

TO

The Memory of

MY FATHER.

'Come, tell us in what place you assemble and gather together your followers?'

PREFECT OF ROME TO JUSTIN MARTYR.

PREFACE.

The history of Christian architecture can be traced in existing monuments as far back as the fourth century. We find it then already represented by imposing buildings, with a distinct plan and an elaborate system of decoration, presupposing a period of tentative efforts. The record of these efforts lies however beneath the surface. The monumental evidence of them is but slight, and it is only by a search into literary records that we can discern their form and character.

It is the object of the present study to bring together this literary material into a shape convenient both for the general reader and the architectural student. The old theory which derived the forms of the Christian church from those of the pagan basilica no longer satisfies the inquirer. The pagan basilica, it is now seen, accounts for several features of the Christian meeting-place, but not for all. For the origin of some of the most important of these we must seek elsewhere, and this search carries us back to the primitive times of the Church. Halls of meeting were used by the Christians of the 'ages of persecution,' long before there can have been any question of copying pagan basilicas, and it is to these that the student of the beginnings of Christian architecture must in the first place direct his attention.

The opening chapter deals accordingly with the general position of the early Christian communities in the Roman world, an understanding of which is necessary before any clear idea can be formed of the probable character of their first places of assembly. We are met here by the theory rendered familiar to the British reader through Dr. Hatch's *Organisation of the Early Christian Churches,* that the Christian communities resembled in certain important outward features the religious associations or clubs of the pagan world. From this the hypothesis naturally follows that the meeting-places of the Christians resembled those of the pagan clubs the technical name of which was *scholæ,* and the reader of these chapters will be asked to consider the view that *the schola of a religious association was the original form of the Christian church.* In the second chapter an attempt is made to illustrate from literary sources the meetings of the primitive Christians, and to convey an idea of the various buildings, *scholæ,* private halls, cemetery chapels, subterranean chambers, in which they were called together.

The remainder of the book deals with the monumental history of ecclesiastical architecture in the West and in the East during the early Christian and Byzantine periods, and with some of the more general questions connected with the history of architecture in the Teutonic kingdoms of the West. For the Appendix has been reserved a more detailed discussion of the relation of the pagan basilica to the Christian church, necessary for the proper understanding of the whole subject, but too archæological in character to be of interest to the general reader.

If the method of treating architectural history pursued

in this book should find favour with the public, it is the intention of the writer to follow it by similar studies on subjects such as ' Monastic Architecture and Art,' and, ' The Ancient Temple.'

The illustrations, as will readily be seen, are merely of the nature of explanatory diagrams, and it was not considered necessary to reproduce any of the familiar woodcuts which make their appearance in so many architectural treatises.

UNIVERSITY OF EDINBURGH,
December 1885.

CONTENTS.

CHAPTER I.

THE CHRISTIAN COMMUNITIES AS RELIGIOUS ASSOCIATIONS UNDER THE ROMAN EMPIRE.

PAGE

The origin of Christian institutions—Connection of the Christians with the Jews, and with other associations of the time—Importance of this in protecting them during the ages of persecution—Influence on Christian forms of the Temple and the Synagogue; the Christians confused with the Jews—First recognition of the Christians as a distinct body in the letter of Pliny to Trajan, c. A.D. 110—Christian meetings connected in this letter with those of clubs and assemblies in general—The associations of the Roman world—The relation to them of the Christians recognised by Tertullian—Its bearing upon the legal position of Christianity—Question of the legal position of the associations in general—Special favour shown by the law to the burial clubs—Their practices illustrated by inscriptions—These practices closely copied by the Christians—The Christian communities as burial societies—Their cemeteries protected, and meetings in them permitted—Special importance of meetings in honour of the martyrs—Christian burial ceremonies—Devotion to the martyrs, and worship at their tombs—The festivals of the martyrs in the ages of peace; their survival in the modern Saints'-Day celebrations, . 1-32

CHAPTER II.

THE EARLIEST CHRISTIAN ASSEMBLIES AND PLACES OF MEETING.

The earliest meetings of the Christians illustrated from the New Testament and from patristic writings—Use of the Synagogues by the Christians in the days of St. Paul—Meetings in private houses—The festal halls of the Roman house—Consecration of them to Christian worship—These meetings cease in the fourth century—Other meeting-places of the Christians during the ages of persecution—(1) *Scholæ*—Form and character of the *Scholæ, Curiæ,* or lodge-rooms used by heathen associations and by the Christians, illustrated from existing remains and inscriptions—(2) *Cellæ* or memorial chapels in the cemeteries used for funeral ceremonies—existing remains at S. Callisto—A memorial cella reconstructed—(3) Subterranean chapels in the cemeteries.

Development from these structures of the Christian Church—Examples of such development from private halls and memorial *Cellæ*—The root-fibres of Christian architecture—Period at which the Christian Church came to assume its regular form—The age of Constantine and the revival of the arts, . . 33-73

CHAPTER III.

THE BASILICA, THE SYNAGOGUE, AND THE CHRISTIAN CHURCH.

Ecclesiastical buildings figured in early Christian mosaics—First appearance of the basilican plan—Connection of Christian architecture with the columned and vaulted

interiors of the ancient world—Origin and architectural development of the basilica, illustrated from Vitruvius and from existing remains—Early use of the basilican form by the Jews—The ancient Jewish synagogue—Its various forms—Description of a normal synagogue of the basilican type—The synagogue as the expression of the national life of the Jews—The Christian basilica of the fourth or fifth century—Its precinct, atrium, and vestibule—Its basilican interior—The terminal apse its important feature—Essential difference between the Christian and the pagan basilica; the latter not apsidal—The private basilica —Summary.
The materials and decoration of the Christian basilica—The art of the catacombs and its distinction from that of the mosaics—Lyric and epic expression—Pictorial decoration in the churches, its origin and intention illustrated from Paulinus of Nola, . 74-134

CHAPTER IV.

THE DOMED CHURCH AND BYZANTINE ART.

Problem of the history of the dome—The Pantheon the earliest monumental dome known—Traces of dome construction on a small scale in Egypt and Assyria—Is the dome Eastern or Roman?—Use of the dome by the Romans, and by the Christians—The earliest domed church built at Antioch—The dome in the East and West in early Christian and mediæval times—Progressive development of dome-construction from the fourth century—Its culmination in Justinian's church of S. Sophia—Description of S. Sophia—Solution in it of problems in dome-construction —Its lighting and decoration—Byzantine culture and art. Later use of the dome—The dome as a Christian form, . 135-178

CHAPTER V.

CHRISTIAN ARCHITECTURE IN THE WEST, FROM THE EARLY
CHRISTIAN TO THE MEDIÆVAL PERIOD.

Description of an early Christian church in the *Apostolical Constitutions*—Evident effort to secure ORDER in church-organisation—Inheritance of this principle from Rome.
Effect on Christian architecture of the Teutonic invasions—Attitude of the barbarians towards Roman culture—Patronage of the Church continued by Teutonic chieftains—Church-building not interfered with—Testimony of Gregory of Tours—The Teutonic genius and Christian architecture—Primitive Celtic and Teutonic building—The 'Roman manner.'
The early Christian church a Schola enlarged, not a basilica simplified—Its development to the Romanesque minster—Successive appearance of Romanesque forms—New movements of the twelfth century—Change in the spirit of architecture—Gothic as the true expression of the Teutonic spirit—Æsthetic character of Gothic—Conclusion, . 179-216

APPENDIX.

NOTE I.
The pagan basilica and the Christian church, . . 217-225

NOTE II.
Sassanid architecture, . . 226-228

LIST OF ILLUSTRATIONS.

FIG.   PAGE

CHRONOLOGICAL TABLE, ... *to face* xvi

Memorial Cella in a Christian Cemetery at the time of the preparation for a Feast in honour of a Martyr, about 270 A.D. Conjectural restoration, . . (*Frontispiece*)

1. Private Houses, from the Capitoline Plan of Rome, . 40
2. 'Insula,' from the Capitoline Plan, . 41
3. Normal Roman House, . . 43
4. Curiæ or Scholæ at Pompeii, from Overbeck, . 52
5. Scholæ (?) from the Capitoline Plan of Rome, . . 53
6. Small Synagogue or Schola, from a mosaic of the Fifth Century, . . . 54
7. Schola beside the Metroon at Ostia, ... 55
8. Plan of Memorial Cella in the Cemetery of S. Callisto, Rome, 56
9. Funeral Triclinium at Pompeii, from Overbeck, . . 58
10. Underground Meeting-place in a Roman Cemetery, . . 60
11. Cella and Basilica of S. Sinforosa, near Rome, from Stevenson, 65
12. Early Christian buildings, from a carved sarcophagus : Fourth century, ... 74
13. Ecclesiastical structures, from a fifth-century mosaic in S. Maria Maggiore at Rome, ... 75
14. Basilica and Baptistery at Ravenna : Sixth century, . 75
15. Section of part of Basilica in its (conjectured) earliest form, . 88
16. Section of part of Basilica. Normal form, . 89
17. Basilica Ulpia from Capitoline Plan, . . 91
18. Basilica Julia, from Dutert, *Le Forum Romain*, . . 92
19. Conjectural Plan of the Basilica at Alexandria, as an example of the Jewish Synagogue of the grandest type, . 107

FIG.		PAGE
20. A Pagan Basilica,		124
21. A Christian Basilica,		125
22. Private Basilica in Palace of Domitian at Rome,		127
23. Model illustrating Dome Construction,		154
24. Model illustrating Dome Construction,		155
25. Ground-Plan of S. Costanza,		157
26. Ground-Plan and Section of S. Lorenzo at Milan,		158
27. Ground-Plan and Section of S. Sophia at Constantinople,		161
28. Plan of a Romanesque church, simplified from that of Hecklingen,		200

CHAPTER I.

THE CHRISTIAN COMMUNITIES AS RELIGIOUS ASSOCIATIONS UNDER THE ROMAN EMPIRE.

In the familiar record of the first days of the Christian Church we read how the men of Galilee, who returned to Jerusalem after the ascension, 'went up into the upper chamber,' which was at once their dwelling-place and their house of prayer and of assembly. There, at the first common meal, the bread was broken and the cup passed round in solemn remembrance of the last occasion on which they had sat at meat with Christ. There, too, they assembled together for their first act of church-government, the election of a successor to the apostate Judas. All is simple and domestic, yet we have here the beginnings of what became in time the most wide-reaching and highly organised of human systems. An elaborate hierarchy, a complicated theology of the sacraments, were to arise out of the informal conclave, the memorial meal; and in like manner, out of the homely meeting-place of the disciples would be developed the costly and beautiful forms of the Christian temple.

It is with the growth and the details of this organised system that the student of Christian antiquities has to deal. His task it is to trace the development of Christianity in its manifold connections as an institution with the world about it, and to attempt to discover the origin and early use of those numerous forms, of which ecclesiastical architecture is one, adopted by it as the apparatus of its outward life. The fundamental fact from which any investigation into these outward forms must start is the following:—that the invention of them is not as a rule to be ascribed to Christianity itself. Christianity, it is true, made certain forms and ordinances its own. At first few and simple, they became in after time very elaborate and numerous; but when their history is pursued backward to its origins, it is found that they were not at first distinctively Christian, but were inherited or borrowed from Jewish sources, or adopted from the practice of other associations, religious and secular, existing at the time in the pagan world.

The hypothesis, that Christianity rather adopted from without than created independently for itself this apparatus of outward life, is one which commends itself to the historic sense of the day,[1] and is in full accordance with the great principle of the economy of causes, which we see at work all about us. It may be argued, indeed, that a special provision of that Divine Providence which guided the fortunes of the Early Church led it to clothe itself in outward forms tending to conceal for a time from public view its unique, original, and, as Hegel has phrased it, revolutionary character. For Christianity was not left to develop in peace and independence. It grew up under the shadow of a colossal despotism, the wielders of which would inevitably regard its claim to universal dominion as insolent defiance. The relation which Christianity was to occupy to the Roman Empire must have been a subject of the most anxious thought to the statesmen of the Early Church, and a dread

[1] The Bampton Lectures of Dr. Hatch, on the *Organisation of the Early Christian Churches*—a work to which the author desires to express his great obligation—may be cited in illustration.

of provoking a conflict explains much of the prudent conservatism of St. Paul on social and political questions. But a conflict more or less open must necessarily arise, and how would Christianity fare? It is here that we find the importance of the conforming tendency just noticed in the Church. The connection of Christianity with Judaism, and with other associations of the time, had a twofold value. If, on the one hand, the Church borrowed largely from these sources both ceremonies and details of organisation, on the other, it owed to them a still more important debt, a debt which it is hardly going too far to call the debt of life itself. During the ages of persecution the Christians were indebted to an incalculable extent to the connections just mentioned. At first they were sheltered through the fact that they were identified by the authorities with the Jews, and the Jews were a privileged sect. Later on, their conformity to many of the proceedings of permitted associations in the Roman world was a source of advantage of a similar kind. The gain in both cases cannot be over-estimated. One cannot speak of the possibility of Christianity passing out of existence altogether under the blows of persecution, but there can be little doubt that, had the gigantic power of the empire been from the first brought steadily to bear against the infant Church, a loss to Christianity only less vital than that of life itself must have ensued. Rome could not have killed Christianity, but she probably could have driven it away from the centres of her civilisation, and in this case the new religion, proscribed at the seat of Empire, would have retired to the East, and have grown up to maturity surrounded by influences tending to develop unduly its mystical and ascetic side. That eminently rational and human character which belongs to Christianity was preserved and strengthened by its position in the very *foci* of the political and intellectual life of the West—at Antioch and Alexandria, at Corinth, Carthage, and Rome. How remarkable is the growth of the community of Roman Christians! Founded perhaps by some of the 'sojourners from Rome' who witnessed the miracle of Pentecost, by the middle of the first century the faith of its members 'is proclaimed throughout the whole world,' and they number among them some who are 'of Cæsar's household.' A little later, in the time of the Flavii, scions of the imperial family itself have embraced the faith, and men of consular rank are ready to lay down their lives for Christ.[1] In the days of Aurelian, the Bishop of Rome already occupies a prominent position in relation to the Church at large.[2] Well did the Church, thus planted and watered at the centre of the world's affairs, justify its privilege. The leaders of its community learned there the lesson which they have never to this day forgotten—the Roman art of ruling men. There they based upon Roman traditions the foundations of a rule more wide and absolute than any which Cæsar dreamed of, and, faithful throughout to the ancient capital, they stood forth amidst the ruins of the Western Empire vigorous in a new life, but encircled still with the prestige of the old. Catching the torch of civilisation from the failing hands of the Cæsars, they bore it forward in the face of the awestruck barbarians, and following still the Imperial motto, 'Incorporate into the State all that anywhere is excellent,'[3] they gathered and united them in a spiritual empire, in which the majesty and policy of Rome were to endure unbroken for another thousand years. The cardinal fact of the Middle Ages—a fact form-

---

[1] *Dict. Christ. Biog.*, artt. 'Clemens Flavius' and 'Domitilla.'

[2] Aurelian (270-275) decrees that certain Church property at Antioch 'shall belong to those to whom the Bishops of Italy and of Rome shall assign it.'—Eusebius, *Hist. Eccl.* vii. c. 30.

[3] See the speech of Claudius to the Roman Senate.—Tacitus, *Ann.* xi. 24, and the original in Gruter, DII.

ing, as we shall see, the basis and explanation of mediæval architecture—is this wide extended dominion of the Romish Church, the foundations of which were laid in centuries when the Emperors, had they so determined, could have driven out the infant community into the desert. That they did not so determine is largely due to the familiar and unobjectionable character of the ordinances adopted by growing Christianity, when it ceased, like 'a new-born babe,' to desire only the sincere milk of the gospel, and came to clothe itself in human institutions, and to take its place as a citizen of the world.

It is with these external relations of Christianity, treated as the necessary basis for a study of early Christian architecture, that the present chapter will be occupied. And first, as to the relation of the early Church to the Jewish social and religious system in the midst of which it arose.

For the present purpose two aspects only of this relation need be noticed, on the one hand the actual derivation of Christian forms from Jewish, on the other the historical connection of Jews with Christians and its influence on the outward history of the Church.

It must in the first place be remembered that the original members of the Christian brotherhood were Jews, and were in no haste to abandon the religious customs of their nation. Christ had come 'not to destroy, but to fulfil,' and the example of the Master strongly inculcated respect for the ancient forms. He frequented with his disciples the modest synagogues of Galilee, remains of some of which have recently been investigated.[1] He went up to Jerusalem at the solemn feasts, and while at Jerusalem 'taught daily in the Temple.' The sanctity possessed by these venerable institutions in the mind of every pious Jew was increased to the disciples by these more dear associa-

[1] Palestine Exploration Fund. Quarterly Statement, 1869.

tions, and when they started forth on their independent career it was with no desire to parade their differences from their brethren of the stock of Abraham. A remarkable illustration of the hold which Judaism retained on even the most liberal of the early Christian leaders is afforded by the story of St. Paul's connection with the four men which had a vow. The incident, as recounted in the twenty-first chapter of the Acts, arose out of a rumour that the Apostle was accustomed to teach the Jews in Gentile cities to 'forsake Moses.' There were at that time (verse 20) thousands of Jews converted to the faith who were at the same time 'zealous for the law,' and in deference to their feelings St. Paul thought it well to follow the advice of his brethren and to give a public proof that he himself also 'walked orderly, keeping the law.'[1] This was not, it must be noticed, an incident of the earliest days of the Church. The Christian preachers had already 'turned the world upside down.'[2] Gentile converts were already numerous, and upon them also 'was poured out the gift of the Holy Ghost.'[3] The hostility of the Jews, both rulers and people, to the new sect had frequently broken out in violent and murderous riots. Yet to the end of his career we find the great Apostle to the Gentiles asserting, both in act and word, his undiminished reverence for 'all things which are according to the law, and which are written in the prophets.'[4] We should naturally, therefore, be prepared to find in Jewish forms the starting-point of the development of those adopted by the Christians.

The religious ordinances of the Jews in the time of our Lord had two aspects. They were in part ceremonial, but in part of a decidedly open and democratic type. The two institutions of the Temple and the Synagogue corre-

[1] Acts xxi. 24.
[2] Ibid. xvii. 6.
[3] Ibid. x. 45.
[4] Ibid. xxiv. 14.

spond to this distinction, and from both of these the Church borrowed largely. The spirit of formalism, of which we see strong signs in some of the first Christian leaders, was nourished from the traditions of the ceremonial law of the Hebrews, and to take one instance, the rite of baptism, appearing in its original simplicity in scenes in the Acts like the impromptu immersion of the Ethiopian eunuch, or the baptism of the household of Cornelius, is already, in the early Patristic writing, the *Teaching of the Twelve Apostles*, encumbered with the Jewish obligation of fasting.[1] The immense importance of this Hebrew rite of baptism, and the various ceremonies involved in it, led to the construction by the Christians of special buildings for its administration, and with these buildings are connected some of the most interesting chapters in the story of ecclesiastical architecture. Another instance in which Jewish temple-ceremonial affected the development of Christian buildings concerns the treatment of the altar. When the Christian Church assumed its distinct form in the third or fourth century, the *presbyterium*, or portion reserved for the clergy and for the altar, had borrowed much of its sanctity from the Holy Place of the Temple of the Hebrews, while equally Jewish was the association of the idea of sacrifice with the Christian altar, in its origin nothing more than a simple table recalling that employed by Christ and His disciples at the Last Supper. Lastly, the sundering of the clergy as a distinct body from the people, which, together with the mystical view of the altar, has had so important an influence upon Christian architecture, is enforced in the Patristic writings by arguments drawn from the Jewish institution of Priests and Levites.[2]

[1] *Teaching, etc.*, vii. 4. Ed. De Romestin, Oxford, 1884.
[2] The influence of the Jewish Temple on Christian architecture is discussed by Weingärtner: *Ursprung und Entwicklung des christlichen Kirchengebäudes.* Leipzig, 1858.

The Synagogue, on the other hand, knowing nothing of Priest and Levite, offered a form of religious worship of a simple and popular kind, familiar and dear to every born Hebrew, and at the same time, it might be thought, exactly suited to Christian needs. The architectural form and arrangement of the Jewish meeting-place are of importance to the present subject, and will be noticed in the following chapters. It is enough to refer here to one or two points of general resemblance between the Hebrew and Christian institutions. In the first place, the habit itself of constant assembly for devotional purposes, at appointed times and places, so characteristic of the Christian communities, is only a continuation of the practice of the synagogue. The Apostolic and Patristic writers exhort the brethren to let their 'assembling together be of frequent occurrence,'[1] by similar arguments to those used in the Talmud to enforce the duty of frequenting the synagogue,[2] and it is the writer of the Epistle to the Hebrews who exhorts his hearers not to forsake the wholesome custom of congregational meeting.[3] The component parts of the service in the synagogue—reading of the Scriptures, singing of psalms, prayer, exhortation—were retained in the churches,[4] though the hierarchical organisation of the latter gave them soon an ordained ministry, and left no room for the pleasant custom of the synagogues, where strangers entering to worship might receive the kindly summons, 'Brethren, if ye have any word of exhortation for the people, say on.'[5] The Synagogue and the Christian Church agreed also in another characteristic. The buildings were by no means only used for Divine service, but were gathering places for the Community, affording opportunity for

[1] Ignatius, *Ep. ad Polyc.* iv.
[2] *E.g.* S. Chrysostom., Migne's ed. i. p. 725. *Babylon Berakhoth*, 6 a, in *Le Talmud de Jérusalem*, translated by M. Schwab, Paris, 1871, etc., vol. i. p. 240.
[3] Heb. x. 25. [4] *Apost. Const.* ii. 57. [5] Acts xiii. 15.

neighbourly intercourse, for teaching and discussion, and also for judicial proceedings. Innumerable stories in the Talmud exhibit to us the Synagogue as the centre of the daily life of the people, in the same way that the early Christian Church was the common resort of the faithful who dwelt in its vicinity. The synagogues, moreover, unlike the Temple and the shrines of the heathen, were often connected with the tombs of the founders, or of local worthies, and we have here a notable point of correspondence with the Christian Church.[1]

It need not be said that this close connection which existed in many respects between Synagogue and Church did not prevent a feeling of intense antagonism between their respective adherents. The first cause of open separation between Christians and Jews was the violent popular onslaught made by the latter upon the followers of Jesus of Nazareth, while Christian apologists, from the author of the *Teaching* downwards,[2] in their bitter criticism on the Jews, laid the foundation for the disgraceful persecutions which stain the history of Christendom. Hence the Synagogue and the Ecclesia were sundered, and their well-known effigies in early Christian art stand apart in antagonism; but it does not follow that the outside observer would at once notice the separation. On the contrary, the connection between Christianity and Judaism was far more patent than the severance which seemed so vital to those actually involved in the controversy, and during the first age of the Church the connection was sufficiently close to make it inevitable that the secular powers should fail to see wherein the difference

[1] See the account of the synagogues of Mesopotamia in the *Itinerary* of Benjamin of Tudela. Ed. Asher, p. 90 ff.

[2] In the *Teaching* it is said (c. viii. 1), that the hypocrites—by whom Ad. Harnack understands the Jews generally—fast on Monday and Thursday, while the Christians have adopted what the writer naïvely believes the more excellent way of mortifying the flesh on Wednesday and Friday.

lay. To Gallio, when he drove his suitors from the judgment-seat, the question at issue seemed only one between two conflicting Jewish sects, and the various officials before whom St. Paul pleaded his cause were inclined to make light of his supposed offence, as being a matter only of Hebrew superstition. It is evident, too, from the testimony of profane literature, that Jew and Christian were, during the first century at any rate, habitually confused.[1] This fact had a twofold result; it was to some extent a cause of suffering to the Christians, but it was also, and to a far greater extent, an advantage, for if it rendered them unpopular, it at the same time secured to their religion and to their society generally a certain amount of legal protection. In his *Antiquities of the Jews*, the historian Josephus rehearses, with patriotic pride, the various decrees of immunity and protection granted in favour of the Jews by the Roman authorities.[2] Pompeius had treated the Jews with brutal severity. Cæsar, on the contrary, had greatly favoured them, and in return for timely assistance they had furnished to him at Alexandria, he granted them certain privileges in consideration of the requirements of their law. They were permitted to 'use the customs of their forefathers in assembling together for sacred and religious purposes,'[3] and to 'make contributions for common suppers and holy festivals.'[4] At Rome, where they had a synagogue, they were patronised by the first Emperors, and soon made themselves of importance. Their legally-recognised position could not preclude a large amount of popular odium, which gathered round them and resulted in the charges of 'hatred to the human race' and the like, familiar to readers of Roman literature; nor did it shield them from occasional measures of severity, as when Claudius 'com-

[1] Such a confusion probably underlies that obscure piece of history, the account of the persecution of the Christians by Nero after the burning of Rome. Merivale, *History of the Romans under the Empire*, c. liv.

[2] *Antiquities*, xiv. 10.   [3] *Ibid.* § 12.   [4] *Ibid.* § 8.

manded all the Jews to depart from Rome.'[1] It gave them, however, an assured status, and the benefit of this assured status was shared, though not with the goodwill of the Jews, by the members of the Christian community. If the Jews were permitted freely to assemble in their synagogues, and carry on there such religious rites as they pleased, Christian gatherings and Christian ordinances had a good chance of escaping hostile notice, and hence it is that Tertullian, in his elaborate survey of the legal position of the Christians in his *Apologeticus*, writes of Christianity as 'standing under the shadow of the illustrious religion of the Jews, a religion undoubtedly allowed by the laws.'[2]

The exact extent to which this semi-legality was secured to Christianity through association with Judaism, and the time through which it operated, it would be difficult clearly to fix. That a period came when Christianity was recognised as an independent system, and was actually placed under the ban of the law, is well known; but it seems probable that the confusion between Christians and Jews continued throughout the first century, for the first time the former are treated as holding an independent position is in the days of Trajan, whose correspondence on this subject with the younger Pliny is of such unique interest. This correspondence brings clearly before us, for the first time, a fact of which there are traces previously, and which was destined, in the periods of active persecution, to protect the Christians in the same way that they had been protected before under the name of the Jews. This fact is *the connection of the Christians with the known and recognised system of association for religious and other purposes prevalent in the ancient world.* The letter of Pliny to Trajan [3] describes the position of Christianity in the province of Bithynia at the opening of the second century, and the relation towards it of the Roman authorities. To

---

[1] Acts xviii. 2.   [2] *Apologeticus*, c. 21.   [3] Plin., *Ep.* x. 96.

such an extent had men fallen away from the old forms of worship that the temples were almost deserted, the sacred rites long disused, and the demand for victims reduced almost to nothing. Information was laid before Pliny, as Roman governor, accusing certain persons of being Christians. Many of these were shown to have been falsely charged, but some confessed the new doctrines. Pliny, who admits that he had had grave doubts whether the mere profession of Christianity was to be regarded as a crime, decides to sentence to death those who persist, on the thoroughly Roman ground that 'whatever the nature of their opinions might be, a contumacious and inflexible obstinacy deserved punishment.' Wishing to know more, he puts two female ministers of the Church (deaconesses) to the torture, but can find out nothing but an 'immoderate and absurd superstition.' All that he could discover about the proceedings of the Christians is the following :—that they were accustomed to meet on a certain day before dawn, and to chant in alternate strains a hymn to Christ as God, binding themselves by an oath—not to any wicked design, but—never to commit either fraud, theft, or adultery, never to break their word, or deny a deposit which they had received. After this they would separate, and later on re-assemble to partake of a simple and harmless meal. This custom, some of the recreant Christians tell him, they had ceased to observe, in consequence of his having issued an edict *forbidding the meeting of clubs and assemblies.*

It is clear from this letter that, in the first place, it was an understood thing that Christianity was illegal, though the exact ground of illegality might be doubtful; and in the second, that it was recognised as coming under the same heading as clubs and assemblies in general, so that a prohibition of these would put meetings of Christians at once under the ban of the law.

Into the nature and the history of these clubs and

assemblies [1] there is no space here to enter. They were voluntary associations formed for mutual benefit, or for accomplishing a common purpose, and resembling in their different kinds the religious brotherhoods, the political clubs, the guilds of workmen, and the friendly and burial societies of the modern world. Their number was so great that at times they literally honeycombed ancient societies, and they were regarded with favour or disfavour by governments according as their religious and charitable character, or the secret political influence which they might be supposed to wield, seemed most prominent. Though some of the societies were politically dangerous, and others cultivated certain foreign forms of religion inimical to good morals, yet, on the whole, their operation must have had a good social effect, especially on the condition of the poor. These friendly associations, indeed, bring out one of the best sides of Paganism. Nowhere would the early Christians have found ideas and habits with which they had more in common than those which were there in force. Liberality of the rich to the poor, mutual assistance, meetings in common, and a common worship—all prevailed in them, and nothing which the pagan world had to show presented more features in common with the Christian brotherhood. What wonder then if the Christians came to conform themselves in large measure to the practice of these *sodalicia*, and so gave cause to the Roman authorities to apply the same laws to both.

The earliest beginning of this connection may perhaps be discerned in the First Epistle to the Corinthians. The picture which the Apostle draws, in the eleventh chapter, of the members of the church assembling themselves together in some place of meeting, which it seems from verses 22 and 34 is not a private house, and there each eating his

[1] Called generally by the Greeks ἑταιρεῖαι ('hetæriæ'), and by the Romans, 'collegia,' 'sodalitates,' or 'sodalicia.'

own supper first, and afterwards sharing with the rest in the celebration of the Communion, agrees in some important respects with the known procedure of the heathen colleges. These had, as one of their institutions, a lodge or meeting-room called '*schola*,' where they held their gatherings, and assembled for common meals, to which each member brought his share of provisions. The words of St. Paul seem to imply the possession by the Corinthian Christians of a place of meeting of a similar kind, and to describe a meal like the above eaten in common, at the close of which the memorial bread and wine would be passed round. The common meal referred to here is the same that St. Jude speaks of under the name ἀγάπη, 'love-feast,' or, in the old translation, 'feast of charity.' Its origin has been variously explained, but it is sufficient to ascribe it to a desire on the part of the disciples, who 'had all things in common,' and ' brake their bread from house to house,' to copy as closely as possible the last evening meal with the Lord. The great popularity which the *agape* attained in the early Church was undoubtedly due in part to the fact that common meals of this kind were familiar institutions throughout the ancient world. They were the most natural form possible for giving effect to a feeling for brotherly union wherever men were brought together by common interests. Not only the pagan 'clubs and assemblies,' but the communities of the Jews, possessed similar institutions, and the law recognised, as we have seen, their periodical contributions and common meals; nor is it likely that the heathen of the time would have perceived any outward difference between their own and the Jewish societies, and those which were addressed by the Apostle in his letter to the Christians of Corinth.[1]

[1] So characteristic of the Christians was this institution of the *agape* that a Latin inscription (Orelli-Henzen, *Inscript. Lat. sel. ampl. collectio*, No. 4073), which records the possession of a cemetery by the 'associates who are accustomed to eat together at a common feast,' may, it is con-

An important passage from Tertullian[1] is the best illustration of this that can be brought forward. He describes the Christians as a body knit together by common religion, by unity of discipline, and a common hope, assembling for prayer, for the reading of the Scriptures, and for the exercise of discipline under the presidency of tried elders. There is a treasure-chest, but 'it is not replenished by purchase-money'—*i.e.* the entrance-fees paid by those who joined the heathen guilds—' as if our religion could be bought.' 'On the monthly collection-day, or whenever he wishes, each one puts in a small donation, but only if he likes, and if he can, for there is no compulsion' (as in the ordinary associations)—' all is voluntary.' The funds are not spent on feasts and drinking-bouts and eating-houses, but 'to support and bury poor people,' and for similar good purposes. The feasts of the Christians are humble; there is none of the extravagance and waste of similar celebrations among the heathen, as at the feasts of the Salii and of the tribes and Curiæ. The name 'love-feasts' explains their nature. Their object is partly to provide sustenance for the needy, who are benefited by the good cheer. The meal begins and ends with prayer. 'As much is eaten as satisfies the cravings of hunger; as much is drunk as benefits the temperate. The talk is that of those who know that the Lord is among their auditors.' In concluding, Tertullian appeals to all whether the assemblies of the

jectured, refer to a community of Christians, who employ this periphrasis to avoid using an unpopular name. The time was not long in coming when the love-feasts tended to become too convivial, and they were separated from the Lord's Supper in the manner described in the letter of Pliny, where the 'simple and harmless meal,' the *agape*, follows the Eucharistic service of the early morning. Later on, they fell into disuse altogether, or only survived in the form of charitable feasts to the poor. Another form of the common meal of the Christians, the memorial feast at the tombs of the martyrs, will be noticed on a subsequent page.

[1] *Apologeticus*, c. 39.

Christians are not innocent and praiseworthy, and entirely unlike those of the secret factions obnoxious to government; or, to quote his own words in c. 38, '*Inter licitas factiones sectam istam deputari oportebat, a qua nihil tale committitur quale de illicitis factionibus timeri solet.*'[1]

The bearing of this upon the legal position of the Christians has now to be considered. We shall have to notice some of the customs of the pagan *collegia* and determine their legality. It will then be necessary to prove that the Christians adopted the same customs, and to show the likelihood that they would in consequence share in the legal protection accorded to their heathen contemporaries. With regard then to the position of the *collegia* in the eye of the law, it has already been mentioned that the government was sometimes friendly, sometimes hostile to them. Trajan was especially severe upon them, and suspected them always of a dangerous political tendency.[2] Alexander Severus and others encouraged them. It follows that under an emperor like the former, all societies of whatever kind would find themselves in danger. Under one like the latter liberty of association would be generally conceded. At the same time there was under every régime alike a recognised difference between legal and illegal associations. No ruler would seek to abolish the institution altogether, while none would grant to it unlimited freedom. It must not be assumed, moreover, that every society had always its distinctly recognised position in one of the two classes. A *sodalitas* might be technically illegal (*non licita*), and yet for various reasons might escape actual molestation. The exact legal position of these societies generally at various periods does not appear yet to have been satisfactorily

---

[1] 'That brotherhood should have a place among law-tolerated societies which commits none of the acts commonly dreaded from societies of the illicit class.' [2] Plin., *Ep.* x. 43.

settled; but we know that on one occasion, when they were suppressed by the Roman government, an exception was made in favour of a class of associations of the poor called *tenuiores*, which, for charitable reasons, were suffered still to exist;[1] and it is argued by some scholars that, under the protection of this general permission, new societies of a similar kind might enrol themselves, and claim a legal authorisation. Now the particular character of these charitable associations lay in the fact that they were largely, if not entirely, burial clubs. No obligation was held more sacred in the ancient world than that of performing all appointed rites to the dead—nothing was so universally respected as a sepulchre. Authorised, therefore, by a decree of the Senate, '*it is permitted to those who desire to make a monthly contribution for funeral expenses to form an association*,'[2] these clubs or colleges collected their subscriptions into a treasure-chest, and out of it provided for the obsequies of deceased members. Funeral ceremonies did not cease when the body or its ashes was laid in the sepulchre. It was the custom to celebrate on the occasion a feast, and to repeat that feast year by year on the birthday of the dead, and on other stated days. For the holding of these feasts as well as for other meetings a building—the above-mentioned *schola*—was necessary, and when the societies received gifts from rich members or patrons, the benefaction frequently took the shape of the presentation of a new lodge-room, or of ground for a new cemetery with a building for meetings. Inscriptions recording these gifts and the thanks of the members benefited are common, and the proofs they give of the mutual good feeling of associates leave a pleasant impression. The following is an interesting example:—

[1] *Corpus Juris Civilis*, Dig. xlvii. 22.
[2] Quoted in the '*Inscriptio Lanuvina*' appended to Mommsen's *De collegiis et sodalitiis Romanorum*, Kiliæ, 1843.

'The place or field which lies on the Appian Way between the second and third milestones . . . on the estate of Julia Monime and her partners, on which there is erected a schola under a portico consecrated to Silvanus and the members of that brotherhood, has been conveyed, free of incumbrance, to the curator and the whole body of members of that brotherhood, for a nominal price, by Julia Monime and her partners . . . that the members may be free to assemble at that place to offer sacrifice, to eat and to make their banquets as long as that brotherhood endures.'[1]

We have recorded here the concession to a pagan brotherhood, whose patron deity was Silvanus, of a space of ground outside the walls, with a lodge erected thereon as a place of meeting. The ground, there can be little doubt, was to be used as a cemetery, and one of the purposes of meeting, as signified in the inscription, was to celebrate the obsequies of members, and to hold memorial banquets as often as their birthdays came round.

So important a matter was it held in ancient times for these memorial banquets and other rites to be continued at the proper anniversaries, that persons of wealth, who perhaps could not trust to the dutiful affection of their surviving relatives, left money to these colleges on the condition that their memory was kept alive by these observances. Numerous inscriptions make mention of this. For example, a college of smiths and carpenters 'erected from the ground at their own expense a new schola upon land presented by Titus Furius.' The same donor left also a sum of money, 'from the interest of which a feast should be held on the anniversary of his birthday.'[2]

The most curious glimpse we have of these funeral customs is afforded by an interesting record of ancient

[1] Orelli-Henzen, No. 4947.   [2] Ibid. No. 4088.

times, which has come down to us in the form of elaborate directions by a pagan Roman of wealth, for the cultivation of his memory, not in connection with a college, but by his family and his freedmen.[1] He provides that his memorial *cella* (chapel) shall be completed according to his plans, and that there be built an *exedra* (alcove) with a seated statue of the testator in marble, and also one in fine bronze, not less than five feet high. Under the alcove is to be a couch with two seats at the sides, all of costly marble. There are to be rugs and carpets kept ready for use on those days on which the *cella* is opened, and two bolsters and pillows, with a couple of dining dresses and two cloaks and tunics. An altar is to be erected in the front to contain his bones. He further provides that all his freedmen and freedwomen, as well as his heir, shall contribute yearly a certain sum, out of which provision is to be made of meat and drink, from which sacrifice is to be offered before the building, and which is then to be 'consumed in that place on the anniversary of my birth, and all are to remain there till the amount contributed has been finished.'

There is something touching in this piece of old-world life brought back to us. Here is a rich man who will have a costly monument with everything of the best—yet marble and fine bronze do not content him. He must know that human lips will speak his name when he is gone, and the gaiety of a human feast echo round his urn. How carefully he provides for the guests at this strange banquet.! All is arranged for them—couches, rugs, cushions, even the festal garments are to be kept in readiness, that nothing be wanting. Yet we note that he has to provide that the feast shall be no mere form. These freedmen and heirs shall not receive the meat and drink and take it home,

[1] See the document given in de Rossi's *Bullettino di Arch. Crist.*, 1863, p. 94.

thinking no more of the dead, but shall remain till all is consumed on the very spot where he is laid.

We turn now from pagan to Christian.

The reader will be prepared for the next step. It is a fact rendered certain by the investigations of the last twenty years that the whole procedure of these funeral colleges, with their contributions, their lodges, their *cellæ*, their meetings, and their burial and memorial feasts, was closely copied, with certain obvious modifications, by the Christians, and it was largely through their adoption of these well-understood and respected customs that they were enabled to hold their meetings and keep together as a corporate body through the stormy times of the second and third centuries. 'It is evident to me,' writes de Rossi,[1] 'that the Christians made use of the privilege of the funeral colleges, and that the Emperors who were well disposed to them, or at any rate not persecutors, availed themselves of this pretext to tolerate the assemblies of the faithful, and to permit the Church to establish itself little by little as a corporation in the bosom of the Empire.' It was as a funeral college that a community of Christians had the best chance of escaping attack, both from the side of the populace and that of the government, and the cemeteries were a secure centre of Christian life at times when in the towns the faithful were, as Tertullian says, 'daily beset by foes, daily betrayed, oftentimes surprised in their meetings and congregations.'[2]

The cemeteries of the Christians existed wherever there was a community of believers, but are chiefly known from the famous examples at Rome which have yielded such rich results to their most recent and indefatigable explorer, Commendatore de Rossi. As we should expect, they were in their origin and many of their arrangements precisely

[1] *Bullettino*, 1864, p. 62. [2] *Apologeticus*, c. 7.

similar to those of the heathen. Ancient cemeteries were always situated outside the walls of the cities, and consisted of a plot of ground (generally bounded on one side by a public road) which was carefully marked out by tablets inscribed with its extent and boundaries, and secured to the possessors by an indefeasible legal title. The ground, or *area*, as it was technically called, was partly laid out in gardens, and partly used for the erection of monuments or of *cellæ*, like the one so carefully described in the Roman testament. Beneath the ground any arrangements desired might be made for the disposal of the remains of the dead. The Christians had, as is well known, their own methods of burial, in the chambers and along the passages hewn at different levels below the surface, but they conformed to the regular custom in having their *areæ* strictly defined, and in keeping their excavations all within the boundaries, until the acquisition of adjoining *areæ* permitted them to extend their operations further. The earliest Christian cemeteries were the property of private individuals, or of Christian families, who offered accommodation in them to the poorer brethren, and the legal title of these possessors was of course as valid as if they had still worshipped the gods of the State. When the Christians came to possess cemeteries in common, they would hold them under the same title as the burial associations, and it was a title which would not lightly be interfered with.

The discussion about the legal position of the Christian communities in the first three centuries turns upon the question whether that recognition by the law, which was certainly accorded to them for some part at least of the time, was granted to them *only in relation to their funeral observances*, or embraced their whole position as religious associations. Whether, before the days of Constantine, the Christians as a corporation could hold a legal title to property is a difficult question;[1] it is held by some that all the direct evidence for the affirmative applies only in strictness to the cemeteries. The crucial moment in the history of the cemeteries is the middle of the third century. Up to that time, except in the case of a few popular raids, the Christian cemeteries and all belonging to them had enjoyed a complete immunity from disturbance. The edict of Valerian, in 257 A.D., for the first time forbade all Christian assemblies and all visits to the cemeteries. It was then that burial-places, or, as they have come to be called, catacombs, became for the first time places of actual hiding, with the ordinary entrances blocked up and access only gained through hidden passages, which opened sometimes into disused quarries and sand-pits. Up to that time the Christians could, as a general rule, perform what funeral ceremonies they liked in the upper air. This was now impossible; and they were forced to construct for their accommodation the subterranean meeting-places or churches which have been met with in some of the catacombs.

The edict of Valerian only remained in force for a brief season, and was succeeded by an important edict of toleration issued by Gallienus in 260, which has been interpreted by some writers as conferring upon Christianity the privilege of a *religio licita*. Nothing more seems, however, to be implied in this edict than the restoration to the Christians of the use of their cemeteries, which had been taken away by Valerian. That the Christians obtained at this time the position of a legal corporation in all its relations is by no means to be inferred. It is true that some evidence that this was the case is to be found in the expressions in the famous edict of Milan, issued by Constantine and Licinius

---

[1] See the article "Christenverfolgungen," by Franz Görres, in Kraus's *Real-Encyklopädie der christlichen Alterthümer*, with the note of the editor on p. 243 ; also Dr. Hatch's Bampton Lectures, 2d ed., p. 152, and *note*.

in 313, as the first charter of the freedom of the Church.
The following words are quoted from the report of the
edict furnished by Lactantius,[1] and refer to the restitution
of church property which had been confiscated under
Diocletian:—'Whereas the Christians are known to have
possessed, besides the places where they were accustomed to
assemble, others also which belonged not to private individuals, but to the whole body of them, that is, of the
Churches;' all these are to be given back 'to the Christians,
that is, to their body and communities' (*Christianis, id est
corpori et conventiculis corum*). These words may be taken
to imply that the Christians had possessed the general rights
of property of a recognised corporation, but it is doubtful
whether anything more is intended by the 'other places'
than the cemeteries. In a letter of Constantine, quoted by
Eusebius, these other places are explained to be 'houses,
fields, and gardens.'[2] The *arcæ* of cemeteries were often, we
know, laid out in gardens, and houses for the caretakers and
sextons were erected upon them, so that there is nothing
here which necessarily proves that the law recognised the
Christians in a wider capacity than that of funeral associations. This is a question which cannot be pursued further.
Whichever way it is finally settled there is no doubt that
the legal title of the Christians to their cemeteries was an
older and a firmer one than any which they may have
acquired in the last half of the third century in relation to
other property, such as land and buildings within the city
walls. If the inscription referred to above (p. 14, *n.* 1), 'LOC[US]
SEP[ULTURÆ] CONVICTOR[UM] QUI UNA EPULO VESCI SOLENT IN
FR[ONTE] P[EDES] ... IN AGR[O] P[EDES] XX,'[3] really refers to
the Christians, it would imply that they held their burial-grounds in the same open way, and by the same tenure, as

[1] *De Mort. Persec.*, c. 48.   [2] *De Vita Const.*, lib. ii. c. 39.
[3] Orelli-Henzen, 4073. See de Rossi, *Bullettino*, 1864, p. 62.

the heathen, who inscribed the fact of their ownership upon the boundary stones of their *areæ*.

The Christian cemeteries on the roads which radiated from the walls of the great cities of the Roman world became in this way the surest and safest rallying-place of the faithful in the ages of the Church's trial. They received a peculiar consecration as the resting-place of the remains of the venerated martyrs.

It is in connection with the funeral rites of the martyrs that we see the full importance to the Christians of these cemeteries, and find at the same time the most curious parallel to classical procedure. All that was done by the pagan associations in regard to the funeral observances and memorial feasts of their patrons was carried out by the Church in connection with the tombs of the honoured patrons of the whole body of the faithful, the saints and martyrs. The earliest evidence of these martyr-festivals is to be found in the *Acts of Martyrdom* of St. Polycarp, A.D. 155, where it is said: 'We took up his bones, more precious than costly jewels, and more highly approved than tried gold, and laid them in a fitting place, where, as far as possible, the Lord will grant us to assemble together with rejoicing and praise to celebrate the birthday of his martyrdom, both in remembrance of those who have fought the fight and for the practice and preparation of those whose time is coming.'[1] So popular did the institution become, and so similar did it appear, to the careless or malicious eye, to the pagan festivals, that an opponent of St. Augustine, in one of his controversies, roundly accuses him on this very ground. 'The sacrifices of the Gentiles,' says Faustus, the Manichæan, 'you change into love-feasts, the idols into martyrs, to whom you pray as they do to their idols. You appease the shades of the departed with wine and food.'

[1] Dressel, *Patrum Apostolicorum Opera*, p. 404.

St. Augustine explains, of course, that the offerings are not to the martyrs but to God, in dutiful remembrance of His servants; but he admits a general similarity in the practices, and himself mentions the memorial feasts in honour of the martyrs, at which, in his time, he confesses, the wine-cup sometimes went round too freely.[1]

For the holding of these funeral ceremonies buildings were needed corresponding to those employed by the heathen societies, and there exists an interesting inscription of the third century which describes the presentation of one to the Christian brotherhood. It was found at Cherchel, in Northern Africa, and is to the following effect:—'A worshipper of the Word (cultor verbi) presented a space of ground for a sepulchre, and out of his own means erected a cella, which he left as a memorial to the holy Church. Hail, brethren, from a pure and guileless heart Euelpius salutes you, O ye who are born of the Holy Spirit!'[2]

We cannot fail to be reminded here of heathen inscriptions, which correspond so closely in wording and in the benefactions described, but lack the bountiful spirit of Christian love which breathes through this. The expression '*cultor verbi*' is noteworthy, for '*cultores*' was the technical term for the members of a pagan religious association in relation to their patron deity. The '*arca*' and the '*cella*' recall the Roman testament before quoted, but how different is the free gift of Euelpius to the brotherhood at large to the self-regarding arrangements of the wealthy Pagan!

[1] S. Augustinus, *Contra Faustum*, xx. 21. Professor Kraus, in the article 'Agapen,' in his *Encyklopädie*, is anxious that the *agape* should be kept distinct from the memorial feast at the tombs. The *agape* was, of course, in its origin a distinct institution, but at the time when memorial feasts were in fashion they were to all intents and purposes *agapæ*, as the words just quoted from St. Augustine's opponent imply. Every common meal of a religious kind for which the Christians came together would be called a love-feast. See also *infra*, p. 59, n. 2.

[2] De Rossi, *Bullettino di Arch. Crist.*, 1864, p. 28.

The funeral ceremonials which were carried on in these *cellæ*, or in times of danger in the crypts below, differed from those of the heathen in the constant reference to the expected resurrection, in accordance with which the burning of the body was forbidden, and the scene relieved from any impression of hopeless gloom. It is to be remarked, however, that the Greek and Roman funeral rites had also a cheerful side, illustrated by the abundant use of flowers, and by the funeral feast in which the departed was supposed in some way to share. There were many funeral customs of the heathen which the Christians could adopt without any sense of unfitness. Among these was the use of flowers and unguents. 'We do not crown the corpse,' writes, it is true, Minutius Felix, in a beautiful passage, 'for the blessed need no flowers, the lost have no joy in them. We twine no fading garland, but we receive from God one blooming with eternal blossom,'—the hope of future blessedness.[1] But Prudentius sings in his funeral hymn how 'the path of light to the broad gardens of the sky lies open now to the faithful soul, while we below with flowers and thick-strewed branches will deck the tomb, and pour the liquid unguent over the inscription and the chilly stones.'[2]

The body, fragrant with embalming spices, was wrapped in fair white linen, or, it might be, dressed in the fine raiment or the official robes which had been the attire of the living. It was then borne to the sepulchre, and, laid in one of the niches or *loculi* of the catacombs, was sealed up till the day of awakening behind a stone, on which a few simple letters witnessed to the name of the deceased, and to the Christian hope. The crumbling bones of the Christians of the early centuries are still to be seen in those strange vaults in the Roman Campagna, whither we descend with the same feel-

---

[1] Minutius Felix, *Octavius*, c. xxxviii.
[2] Prudentius, *Cathemerinon*, x. v. 161 *seq.*

ings of interest and awe which, more than fifteen hundred years ago, impressed themselves on the youthful Jerome when, as he tells us, he was wont on the Sundays to go round with his companions to visit the sepulchres of the martyrs.[1]

De Rossi describes, as an eye-witness, the discovery of more than one body which, embalmed with spices, and enclosed in a marble sarcophagus, had remained in a wonderful degree of preservation. One that he himself assisted to move was that of a lady, the golden ornaments of whose robes were still to be seen, while the crushed bones of the skull and the marks of blood seemed to betoken martyrdom.[2]

A still more extraordinary example of preservation is that of the body which has been connected with the name of one of the most famous of the early Roman martyrs, St. Cecilia. A body, presumed to be that of the saint, and unquestionably that of an early martyr, had been conveyed from the catacombs by Pope Paschal I. in 821, and placed beneath the high altar of the Church of S. Cecilia in Trastevere at Rome. There, in the year 1599, it was discovered and brought to light, and its appearance has been minutely described by the famous Cardinal Baronius, and the antiquarian Bosio. 'I saw,' writes Baronius,[3] 'enclosed in a marble sarcophagus, a coffer of cypress-wood, which enshrined the sacred limbs of Cæcilia. It was covered with a sliding lid, a little decayed. . . . Within we found the holy body of Cæcilia laid as it had been found by Pope Paschal, with the veil steeped in her blood at her feet. Some threads of silk embroidered with gold, which were visible, were the remnants of that gold-enwoven robe mentioned by Paschal, which was now decayed by age. Another vesture of silk, of light texture, laid over the body of the martyr,

---

[1] S. Hieron., *in Ezech.* xl., Migne's ed., vol. v. p. 375.
[2] De Rossi, *Roma Sotterranea*, vol. ii. p. 125.
[3] *Ad annum* J.C. 821, § 15.

and clinging to it, permitted the posture and the form of the limbs to be seen. It was remarkable then to behold how the body was not lying supine, as if in a tomb, but as a maiden might lie on her couch, upon the right side, with the knees a little drawn up, looking more like the form of one who slept than of the dead, and so ordered as to convey to all beholders the idea of virgin modesty.' The beautiful statue by Stefano Maderna, which is to be seen in the church, is said to reproduce with accuracy the attitude of the body as it lay in its coffin of cypress-wood.

Brought thus into personal contact, as it were, with Christians of the age of persecution, we can realise more readily the circumstances under which they were laid to rest, and sympathise even at this distance of time with the feelings of the survivors. The sentiment of the early Christians for their martyrs was of the intensest kind. On the one hand, the thought of them called out the most enthusiastic feelings of personal devotion; on the other, they appeared before the faithful as standing witnesses to the most momentous articles of their belief. Christianity stands alone among religions in evoking that passion of personal affection, which is the strongest feeling in human nature. Devotion to the martyrs fed this passion till it carried the faithful away into extravagances which the Church authorities found it necessary to control. Such were the honours paid to any who suffered—not the pains of death—but even the slight inconvenience of imprisonment for the faith, that interested persons took advantage of the liberality of the faithful, and the satirist Lucian humorously describes how a rogue of his acquaintance, at his wits' end for means, at last turned Christian, had himself imprisoned, and was immediately fêted as a hero.[1]

---

[1] The passage, an amusing one, is in the dialogue called 'Peregrinus,' § 12, and runs as follows :—' Proteus is now thrown into bonds, but this very

This is, of course, merely the excess of a natural and noble enthusiasm. It had deeper roots than mere hero-worship. The martyrs were, in the estimation of the early Church, the first-fruits of the sacrifice of Christ. They were the witnesses of that faith in the resurrection of the dead, for which Paul had been called in question. 'They have persuaded themselves that they are immortal,' writes Lucian, 'so they despise death, and seek it even of their own free will.'[1] There was nothing in which the Christians differed from the adherents of the heathen creeds more vitally than in this belief in a life of personal activity beyond the grave. It is true that under the Roman Empire the faith in the old gods of the Greco-Roman theology had become a thing of the past. All that was religious in the Paganism of the second and third centuries centred in the foreign cults of Isis and Serapis, of the Magna Dea, and of Mithras. These oriental religions, with their mystic rites, in which men could 'approach the confines of death, and passing the threshold of Proserpina, penetrate to the hidden source of all things,' in which they 'saw, at midnight, the sun blazing in clearest radiance, and entering the presence-chamber of the infernal and the heavenly gods, adored them face to face,'[2] might answer in some measure to the craving for something

circumstance won for him no small reputation, starting him afresh in life, and ministering to his delight in notoriety and sensation. For when he was put in prison the Christians took it greatly to heart, and moved heaven and earth to get him out again. When this could not be done, they set to work in right good earnest to do everything they could to serve him, and from early dawn you might have seen ancient widows and orphan children standing about the prison, while those in authority bribed the gaolers, and got themselves in to pass the night with him. All sorts of dainties were then introduced at meal-time, and sacred discourses held, in which he won the name of a new Socrates. Incredible is the fuss made when the Christians take up publicly a matter like this, and no expense is spared. Hence Peregrinus made himself master of no small sums of money, and gained considerable profit through his bonds.'

[1] Lucian, *De morte Peregrini*, § 13.   [2] Apuleius, *Metam.* xi. 23.

deeper and more inward than the ancient State religions had to offer. But how different were the dim glimpses of the unseen which their votaries thought to gain in initiation, to that clear vision, that inspiring hope, which made the light of the Christian's life! Through the martyrs men felt that they came nearer to the unseen world, into which these had entered, and on the threshold of which the survivors might at any moment be standing. Hence, when the waves of persecution were closing around, the Christians clung more and more closely to the assurance which these examples offered. To realise the feelings of those times we must transport ourselves in thought to the Christian cemeteries when, in the days of sorest persecution, they afforded the only hope of a safe meeting-place. Thither, in the dusk of twilight, by unfrequented ways, in twos and threes, we can picture to ourselves the Christians gathering. Entering by a secret passage the closed, and perhaps guarded, catacomb, they assembled in the subterranean chapels, where the bodies of the martyrs lay at rest. With the sarcophagus for an altar, the holy ceremonies were performed, and perhaps a hurried love-feast eaten in memory of the dead. There, while the paved road without might be ringing with the tramp of the cohorts of the watch, the spirits of the martyrs would seem to come near to them, and inspire them to dare all for Christ. There even, as the priest laid his hands on the bread of the communion, they might behold in their ecstasy beneath the altar, 'the souls of them that had been slain for the word of God,' and hear them cry 'with a great voice, saying, How long, O Master, the holy and true, dost thou not judge and avenge our blood?' till from all the congregation is breathed back with one accord the words of answer: 'Tarry yet a while till we also and our brethren, who shall be slain even as ye were, shall be fulfilled.'[1]

[1] Rev. vi. 9-11.

The popular enthusiasm which attached itself in the times of persecution to the martyrs' graves, found new methods of expression in the age of the Church's triumph. The anniversaries of the martyrdom of the saints, or, as they were called in imitation of the pagan custom, their *birthdays*, became the occasions of great popular festivals, like sacred Dionysia or civic games, with which indeed Church writers compare them. The situation of the cemeteries outside the walls was a favourable circumstance, and made the fête a country one; so that St. Chrysostom contrasts the town festival of the Maccabees at Antioch, when the rustics all thronged in through the gates, with the rural feasts of the martyrs, which left the towns deserted.[1] On such occasions St. Jerome proudly boasts, 'the gilded Capitol is horrid with neglect; with smoke-stains and spiders' webs are covered all the temples of Rome, the city is moved from its seat, while the torrent of the populace sweeping past the ruined shrines pours forth to the martyrs' tombs.'[2]

It would be easy to multiply quotations to illustrate these festivals. The literature of the time is full of notices of them, and we see how they rapidly came to take the position of Christian substitutes for the frequent pagan celebrations which had come to be considered so indispensable. The excesses which are wont to accompany popular festivities on a large scale were not wholly wanting here, and a passage from St. Augustine has already been mentioned (page 25), which shows the tendency of the memorial feasts to become mere convivial gatherings, wherein, to borrow the sarcasm of St. Jerome, 'the martyrs, who had pleased God by fasting, were themselves honoured by surfeiting and drunkenness.'[3] It is not to be wondered at that

[1] S. Chrys., *De sanctis Martyribus*, § 1. Migne's ed., vol. ii. p. 647.
[2] S. Hieron., *Epistolæ*, cvii. 1. Migne's ed., vol. i. p. 868.
[3] *Ibid.* xxxi. 3. Migne, *l.c.*, p. 446.

Church writers and Church Councils felt it necessary, even in the fourth century, to put a curb upon the practice.

The time was soon to come, however, when these expeditions and rural festivals were to be cut short in another fashion. The incursions of the barbarian and the infidel rendered the country unsafe, and confined the population within the walls of the cities. Italy became the scene of incessant invasions and counter invasions, and the inroads of the Goths and of the Lombards reduced the Roman population to the utmost straits, while the barbarians even plundered the cemeteries themselves. It was in these times that the practice was adopted of conveying for safety within the city walls the priceless treasures of the catacombs, and an inscription in S. Prassede records how, in 817, Pope Paschal translated into Rome no fewer than 2300 bodies of the saints, while cartloads of relics are said to have been buried at the same period in the Pantheon, or, as it was called by the Church, S. Maria ad Martyres. The arrangements which were made in the churches for the reception and conservation of these relics had an important bearing upon architecture, which will be noticed in the next chapter. The festivals of the martyrs still continued to be held in the places to which their bones had been removed, and led on by insensible transitions to the modern saints-day celebrations. The semi-pagan character, which still belongs to many of these 'festas' in Southern Europe, has often been commented on, and, even in the heart of Protestantism, the Public-School boy of to-day, who enjoys his half-holiday on the festival of a saint, is perpetuating an institution of the early Church, which, itself resting on classical tradition, is linked on to the joyous popular feasts that he reads of in Herodotus or Plato.

CHAPTER II.

THE EARLIEST CHRISTIAN ASSEMBLIES AND PLACES OF MEETING.

FROM the general position of the Christian associations in the Roman world of the first three centuries, we pass on to the question of the form and character of their earliest meeting-places. Little aid can here be derived from a study of existing monuments. The earliest monuments of Christian architecture of which we possess actual remains are the basilicas and domed churches of the fourth and fifth centuries at Rome, Ravenna, and Milan, in central Syria and in northern Africa. The aspect of these corresponds to an advanced stage of church organisation, and to a condition of worldly prosperity which did not exist until after the age of Constantine. The primitive communities would, we may be sure, have cared little about the situation, the form, and the decoration of their meeting-places, but the numerous Syrian churches of about the fifth century, described by de Vogüé,[1] are solidly built and exceedingly conspicuous; S. Lorenzo at Milan, from the end of the fourth, shows us an architectural conception of the utmost boldness carried out with accomplished science, while in the basilicas of the same period at Rome and Ravenna we find a regularly ordered plan, with distinct places for different sections of the community, with ecclesiastical fittings to supply the requirements of ritual, and with splendid decoration and didactic pictures to reflect the glory of the Church, and

[1] *Syrie Centrale, Architecture civile et religieuse du 1ᵉʳ au 7ᵉ siècle* Paris, 1865.

bring home its history and its doctrines to the mind of the convert.[1]

These existing examples, together with those described to us by writers of the period, convey an idea of Christian architecture very different from the reality of the primitive age, and it would be a great mistake to argue back from them to the original forms. For the earlier period we are forced to rely upon literary records, illustrated by what we know about the non-Christian buildings which are likely to have served as models for the meeting-places of the faithful.

The New Testament contains less information than might have been expected about Christian meeting-places, for the word 'ecclesia,' though commonly used in after time to denote a *place of assembly*, seems in the Acts and Epistles to have no local significance, but to mean only the *congregation of the faithful*. One passage, however, in which distinct reference is made to a regular place of Christian worship occurs in the Epistle of James, and here the word employed is 'synagogue.'[2]

The use of this word would suggest that it is to the Synagogue that we must look for the pattern of the first Christian meeting-house, but we might easily fall here into a mistake as great as that which has just been guarded against. The Jewish synagogues, like the basilicas of the fourth century, were often large and imposing structures, and it is obvious that the earliest Christians can neither have desired nor been able to erect buildings of the character of these legally sanctioned edifices, the outcome of a

---

[1] The early mosaic pictures in the churches of Rome and Ravenna contain contemporary representations of these buildings and of their interior fittings, which supplement the literary accounts, and form an indispensable groundwork for a discussion of the church architecture of the time.

[2] James ii. 2, 'If there come into your synagogue a man with a gold ring, in fine clothing . . . and ye have regard to him that weareth the fine clothing, and say, Sit thou here in a good place,' etc.

time-honoured and elaborate religious system. The fitting place, therefore, to discuss the form and arrangement of the synagogue will be in connection with Christian buildings of the more advanced period, and this must be deferred till the next chapter. The early use of the synagogue by the Christians, before their final separation from the Jews, is, however, a point of some interest, which a moment's attention to some familiar passages in the Acts will enable us to elucidate.

Now it is clear from the New Testament that as the seer of Patmos 'saw no temple' in the city which descended before his eyes from heaven, so no idea of a specially Christian form of building occurred to the members of the primitive Church. The Temple at Jerusalem was still sacred to them. Thither they still ascended at the hour of prayer,[1] and one of its porticoes they chose for a time as their place of daily assembly.[2] Away from the capital they still followed the customs of their people, frequenting the synagogues and places of prayer, and freely delivering therein some of the first Christian sermons.[3] It was the more easy to propagate the new doctrines in this manner through the fact that, as we see in the accounts of the proceedings of St. Paul on his missionary journeys, the congregation in the synagogues was of a mixed description. The cities visited by him, Antioch, Iconium, Philippi, Thessalonica, though they might contain a large Hebrew population, were not Jewish, but Hellenistic. So attractive to the pagan mind at this period was the religion of the Jews[4] that the Greek or semi-Greek inhabi-

---

[1] Acts iii. 1.   [2] *Ibid.* v. 12 ; Luke xxiv. 53.
[3] Acts ix. 20; xiii. 5, 16 ; xiv. 1, etc.
[4] A striking example of this is the case of the centurion at Capernaum (Luke vii. 5), who need not have been a regular proselyte. Josephus (*Wars*, ii. 20, § 2) mentions that almost all the matrons of Damascus were addicted to the Jewish religion. Horace, Juvenal, and Tacitus testify to the influence of the Jews in the Roman world.

tants of these cities seem to have formed the habit of attendance at the synagogue, and we find the Apostle addressing his audience by such terms as 'Men of Israel, *and ye that fear God*,' 'Brethren, children of the stock of Abraham, *and those among you that fear God*.'[1] So successful was this appeal at Antioch in Pisidia that 'the next Sabbath *almost the whole city* was gathered together to hear the Word of God,'[2] and Paul turns from the Jewish to the Gentile section of his hearers as if both had an equal right to be present.[3] At Iconium 'they entered into the synagogue of the Jews, and so spake, that a great multitude *both of Jews and of Greeks* believed.'[4] At Corinth he 'reasoned in the synagogue every Sabbath, and *persuaded Jews and Greeks*.'[5]

It is in accordance with those open and democratic characteristics of the worship of the Jews already referred to[6]—characteristics expressed by the sacred motto, 'Mine house shall be called an house of prayer for all peoples'[7]— that we find it the custom for any suitable person to lead the service and offer the word of exhortation in the synagogues, and for any who chose to enter and share in the ministrations provided. Nor did the controversial character of the addresses delivered by the Apostle make them unsuitable for the place or the audience. Either the synagogue itself, or a side building closely connected with it, was the recognised place for rabbinical disputations, and these we know were sometimes carried on, as by the rival schools of Hillel and Shammai, with no little vehemence. The brilliant and zealous disciple of Gamaliel I. would be certain to attract numerous hearers, and the interest of his speech might to the majority of the mixed audience excuse its want of orthodoxy. In other cases earnest Jews, as at

---

[1] Acts xiii. 16, 26. [2] *Ibid.* 44. [3] *Ibid.* 46, 48.
[4] *Ibid.* xiv. 1. [5] *Ibid.* xviii. 4. [6] *Ante*, p. 8.
[7] Isaiah lvi. 7.

Berœa,[1] would carefully examine, before condemning, the arguments, so that the Apostle could speak boldly in the synagogue at Ephesus 'for the space of three months, reasoning and persuading as to the things concerning the kingdom of God.'[2]

We come now, however, to the point when we must for a time turn our back upon the synagogue, for the verse following that just quoted contains a notice of great significance for the period on which we are about to enter. During these three months it appears that the Christian converts assembled with the other Jews in the regular synagogue where Paul was accustomed to offer exhortation, and no division seemed to either party to be called for. At the end of this time occurred a manifestation of popular hostility similar to those which had so often cut short at other places the Apostle's ministry, and in consequence, we are told, he 'departed from them,' ceasing to frequent the synagogue and 'separated the disciples.'[3] The next step was not, as might perhaps have been expected, to formally open a new 'synagogue of the Christians.' The new community had, as we have seen, neither the worldly means nor the legal position which would have enabled it to erect for its accommodation imposing public buildings like the synagogues. It had to have recourse to interiors of a humbler and more private kind, and it is with these that the story of Christian architecture really begins.

Following then the Apostle's steps at Ephesus, we find him leading the disciples to the 'School of Tyrannus.'[4] Tyrannus may have been a Christianised Jewish instructor, or a Grecian Sophist who was either a convert or was willing to let out his lecture-hall to the Christians—in any case his room was a private one, and meetings held there would have the same informal character as those in the house of

[1] Acts xvii. 11.   [2] *Ibid.* xix. 8.   [3] *Ibid.* 9.   [4] *Ibid. l.c.*

Titus Justus, hard by the synagogue at Corinth, in which the Apostle had previously, under similar circumstances, taken refuge.[1]

It seems clear that the first meeting of the Christians as a community apart—meetings, that is, of a private rather than a proselytising character—took place, as we see from Acts i. 13-15, in private apartments, the upper rooms or large guest-chambers in the houses of individual members. Such a room was doubtless provided by the liberality of Titus Justus, such a room again was the upper chamber in which St. Paul preached at Troas;[2] in such assembled the converts saluted by the Apostle as the 'church which is in the house' of Aquila and Prisca,[3] of Nymphas,[4] and of Philemon;[5] while it was in his own hired dwelling that at the close of his career he received and addressed his countrymen at Rome,[6] who must often have united with him there at a love-feast or eucharistic service.

In the literature of the age which succeeded to the apostolic, there is abundant evidence of the continuance of this practice. Meetings of the Christians at Rome in private houses are referred to in the *Acts of Martyrdom* of Justin Martyr and his companions, in the second century,—a document which is probably based upon the official notes of the trial itself. The Christians were brought up before the Prefect of Rome, who asked them, 'Where do you assemble?' Justin said, 'Where each one chooses and can; do you suppose that we are all accustomed to meet together in one place? Quite otherwise, for the God of the Christians is not confined by place, but being invisible, He fills the heaven and the earth, and the faithful everywhere adore Him, and sing His praise.' Then the Prefect, 'Come, tell us in what place you assemble and gather together your

---

[1] Acts xviii. 7. [2] *Ibid.* xx. 7, 8. [3] 1 Cor. xvi. 19.
[4] Col. iv. 15.   [5] Phile. 2.   [6] Acts xxviii. 30.

followers?' Justin answered, 'I have lodged up to this time by the house of one Martinus, at the bath which is called Timiotinian. This is the second time that I have been in Rome, and I know no other place but the one I have named. There, if any one wished to come to visit me, I communicated to him the Word of Truth.'[1]

These meetings in private houses, though they belong to the earliest, and, as we might say, the domestic age of the Church, by no means passed out of use when Christianity became a more public institution. Until freedom of religion was proclaimed by Constantine, and the permission which the Church obtained to receive legacies poured into her lap abundant funds for church building, the custom of meeting in private houses seems still to have remained in vogue. That there was nothing in this custom incompatible with the position of the Christian communities as *religious colleges* is proved by heathen parallels. Tacitus speaks of associations devoted to the worship of Augustus which 'held meetings after the manner of the colleges from house to house,'[2] and an inscription tells of 'the college of the elders and of the juniors in the house of Sergia Paullina.'[3] It becomes, accordingly, a problem of interest in the history of Christian architecture to determine what was the form and character of these primitive meeting-places, which may have contributed to form in the minds of the earliest Christians their nascent idea of an architectural style suited to their needs. A word as to the arrangement of antique dwelling-houses will here be necessary. These may be divided into two kinds—the normal house built where there was freedom of space, and the city residence amidst crowded streets. The former had the principal rooms on the ground-floor, and covered in consequence a large area in proportion to its

---

[1] Ruinart, *Acta Martyrum Sincera*, Ratisbon, 1859, p. 106.
[2] Tacitus, *Annales*, i. c. 73.   [3] Orelli-Henzen, 4938.

height. Most of the houses laid bare at Pompeii are of this kind, and on the fragments of the ancient Capitoline plan of Rome, which dates from the time of Septimius Severus, there are several of a similar form indicated (Fig. 1),[1] and these will presently be described.

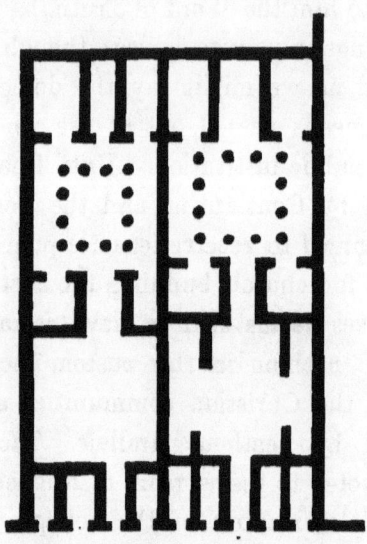

FIG. 1.—Private Houses from the Capitoline plan of Rome.

The primitive Roman house had only one story, but as cities grew to be more densely populated upper stories came into use, and it was the custom to place in these the dining apartments, which were called *cenacula*. Such apartments would answer to the 'upper rooms' mentioned in the Scriptures in connection with Eastern houses, and associated with the early days of Christianity. The immense increase in the population of Rome under the early emperors, of which poets and satirists so feelingly complain, made building land very costly, and to gain room story upon story was added to the houses, until considerations of safety obliged the government to fix a maximum of height. Martial tells us that he lived up three pair of stairs, but they were very

[1] Jordan, *Forma Urbis Romæ*, frag. 172.

high ones.[1] These lofty houses were built in blocks, called *insulæ*, and let out to numerous tenants, who paid very high rents for even the uppermost floors. It is probable that Fig. 2,[2] from a fragment of the Capitoline plan, represents a block of this kind, the shops which surrounded it on

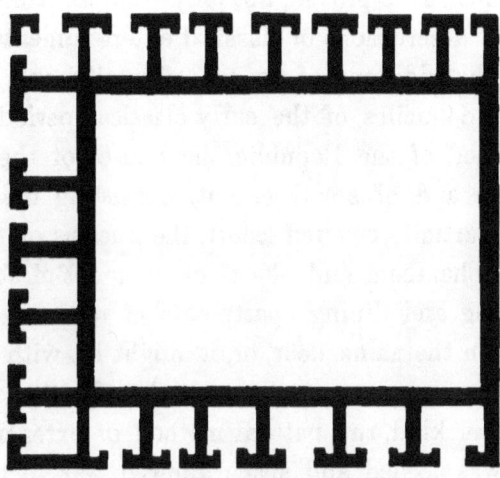

Fig. 2.—'Insula,' from the Capitoline plan.

the ground floor being alone marked in detail. The Martinus mentioned by Justin may have owned an *insula*, or a portion of one, and let out rooms which were used by the Christians as a place of assembly. In the anti-christian dialogue *Philopatris*, formerly ascribed to Lucian, we are introduced to a Christian gathering in an upper chamber of one of these city houses. The speaker describes how he was taken up a long winding staircase and ushered into a handsome room with a gilded ceiling, wherein was seated a company of men of a gloomy aspect, who were gloating over the misfortunes impending on the human race!

[1] Strabo (xvi. 11, § 13) writes of Aradus in Syria, that even in his time 'the population was so large that the houses had many stories;' and of Tyre (*ibid.* § 23), that 'the houses consist of many stories, of more even than at Rome.'

[2] Jordan, *Forma Urbis Romæ*, frag. 191.

Of more importance in relation to the beginnings of Christian architecture are the meetings of the Christians held in private mansions of the normal kind, which differed entirely in plan from these many-storied residences in the crowded parts of the cities. The existence of capacious halls in these mansions is proved, not only from existing remains, but from the descriptions of classical entertainments. These halls were the additions of an age of wealth and luxury to the modest domiciles of the early classical period. Up to the last period of the Republic, the houses of the Romans were simple and of small extent, consisting only of the *atrium*, or partially covered court, the nucleus of the home, some small chambers and offices opening out of it, and one or two living and dining apartments of moderate size beyond,—all on the same floor, or, it might be, with the addition of some upper rooms or *cenacula*. In the case of a house of this kind, the natural method of extension, when extension was needed and space allowed, was to throw out new apartments at the back, where land could most easily be acquired. These adjuncts were built in the Greek style, and called by Greek names. They consisted of a spacious court or *peristyle*, surrounded by a colonnade out of which opened sundry festal apartments, called *oeci*, from the Greek word οἶκος, a house or hall, and varying in number size and splendour with the wealth and position of the owner. A garden often lay beyond, into which the *oecus* opened.

Fig. 3 gives the normal plan of a Roman house of the later classical period, with the *atrium* and surrounding apartments, and the *peristyle* with festal halls and garden beyond. In the court of the *peristyle* would be found a fountain, with statues, flowering shrubs, etc.

The Greek house received similar additions in the post-Alexandrine period of private luxury, and it was in the apartments of this portion of the establishment that were held the entertainments described by writers of the later classical epoch. The Christian communities contained from an early period members of wealth and social position, who could accommodate in their houses large gatherings of the faithful; and it is interesting to reflect that while some of the mansions of an ancient city might be witnessing, in suppers of a Trimalchio or a Virro, scenes more revolting to modern taste than almost anything presented by the pagan world, others, perhaps in the same street, might be the seat of Christian worship or of the simple Christian meal.

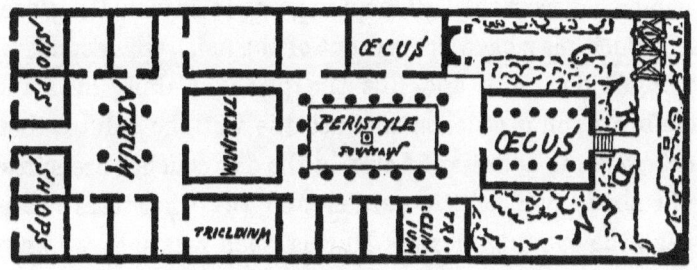

FIG. 3.—Normal Roman House.

The architectural style of these halls varied, and Vitruvius mentions four kinds,[1] though his words are hardly clear enough to afford grounds for a reconstruction. Upon this point something more will be said in the next chapter, but it may be remarked here that the richness of decoration of these festal halls and *cenacula* must have had a potent effect in preparing the Christians to apply similar adornments to their regular churches, as soon as these came to be built. It is with the employment of these structures, rather than with their architecture, that we are at present concerned. The use of these large private halls for Christian gatherings is well illustrated in the curious theological romance known as the *Recognitions* of Clement. The work professes to describe the proceedings of St. Peter and a band of his followers

[1] *De Architectura*, vi. 3.

in the course of a missionary journey near Palestine, during which discourses are preached and conversions made without number, and a brisk controversy kept up with 'Simon Magus.' On one occasion the party comes to Tripolis and is lodged in the house of one Maro. In the morning a concourse of those eager to hear the word assembles before the gates. Peter asks where there is a suitable place for discussion. Maro replies, 'I have a very spacious hall capable of holding more than 500 persons; there is a garden too within the house.' Peter said, 'Show me the hall and the garden.' When he had seen the hall, he was going in to view the garden also (which may have opened out of the hall), when suddenly the crowd rushes in and fills the place. Taking his stand, therefore, upon a pedestal against the wall, he salutes them and begins his address. Later in the day couches are spread in the shady part of the garden, and the Christians recline to take the evening meal.[1] During such meals, as we know from other sources, the younger brethren who had good voices would lift them up in psalms.

At El Barah, in central Syria, not a long distance from Tripolis, de Vogüé found the well-preserved ruins of a handsome country villa, surrounded by a large garden, and containing a hall of fine proportions, from which we may derive an idea of this of Maro.[2]

Another passage in the *Recognitions* is interesting as a record of the formal assignment, by a wealthy convert, of an apartment in his mansion for the purposes of Christian assembly. At the close of St. Peter's mission in Antioch, so runs the romance, 'a certain rich man of the place, Theophilus, who was more exalted than all the nobles of the country, consecrated the immense hall (basilica) of his palace as a Christian church, and there, by the acclamation of the

---

[1] *Clementine Recognitions*, iv. 6.
[2] De Vogüé, *Syrie Centrale*, pl. 51-53.

people, was set up the chair of the Apostle Peter.'[1] Tradition makes this Theophilus afterwards bishop of Antioch, in the same manner as in the *Recognitions* Maro, the owner of the hall and garden already mentioned, is made by St. Peter bishop of Tripolis. In this way not the hall alone, but the whole residence, might be converted to Christian uses, and become the recognised centre of Christian life and work for the town and the surrounding country. We may be permitted to conjecture, indeed, that it was a maxim of church policy to secure as bishops men of wealth and social station, in whose houses the adherents of the faith should have an advantageous position from which to extend their operations. Such clergy-houses—if the expression may be used—might easily be transformed a little later into monasteries. The arrangement of the mediæval monastery recalls that of the ancient house. The cloistered court is the *peristyle;* the church, the refectory, the chapter-house, and other apartments round it correspond to the *oeci* of classical times, and the resemblance may be due to this derivation.

A Christian establishment of the kind just noticed we have brought before us in a curious document[2] referring to the time of the Diocletian persecution, 303 A.D., and recording the official visit of the Roman magistrate of the town of Cirta, in Africa, to the domicile of the Christians, in quest of their sacred writings which had been confiscated by Imperial edict. 'When they came to the house where the Christians were wont to assemble, Felix, *flamen perpetuus* curator of the colony of Cirta, said to Paul the bishop, "produce the writings of the law and anything else which you may have."' The bishop fenced with the demand, and said that 'the readers had the copies of the Scriptures,' but a search

---

[1] *Clementine Recognitions,* x. 71.
[2] Known as *Gesta Purgationis Cœciliani,* and printed in the Appendix to vol. ix. of Migne's edition of St. Augustine's Works, p. 794.

was made, which is described as follows:—'There was seated Paul the bishop, with Montanus and Victor, Densatelius and Memorius, presbyters; standing by were Martis, with Helius, deacons; Marcuclius, Catullinus, Silvanus, and Carosus, subdeacons; Januarius, Meraclus, Fructuosus, Migginis, Saturninus, Victor, and others, sextons. Victor, the son of Aufidius, took down the inventory, as follows:—Two golden cups, item, six silver do., six silver ewers, a silver vase, seven silver lamps, two candelabra with branches, seven short bronze candlesticks with their stands, item, eleven bronze lamps with their chains; 82 women's tunics, 32 capes, 16 men's tunics, 13 pairs of men's shoes, 47 pairs of women's do. . . . Afterwards search was made in the library, but the cases were found empty. . . . When the dining-room (*triclinium*) was opened there were found there four jars and six vases. . . . The search was concluded by a visit to the houses of the readers, where various copies of the Scriptures were given up to the authorities.

The scene here is undoubtedly a private house, the residence of the bishop, but devoted to Christian purposes. The hall is the regular place of meeting, and here the bishop sits in state to receive the magistrate, with two presbyters on each side of him. The deacons and under officers stand at the side. A chamber near, or a recess in the hall, is the treasury, and here are found the various objects used at the gatherings. The library, which was always a feature in houses of the wealthy, is hard by, with its cases for books. The dining-room is the place where the love-feasts are celebrated. The inventory of objects found is curious. The lamps and candelabra which figure so largely show that meetings were held in the evening or before dawn. The cups or chalices and the ewers were used at the celebration of the Eucharist. The jars and vases found in the *triclinium* may have been employed for the purpose of measuring out the portions of wine and oil assigned at the love-feasts to the poorer members. The most singular item is the abundant supply of men's and women's garments kept in store. The custom of putting on a festal robe for dining is illustrated by the accounts we possess of classical banquets, as well as by the parable in the Gospel of the feast and the guest who had not on a wedding garment. It is possible that these robes were used for the attiring of the poorer brethren at the love-feasts, that no one of the faithful should have cause to be ashamed of the poorness of his habit. We are forcibly reminded of the similar provision of robes for the memorial banquet referred to in the Roman testament quoted on an earlier page.[1]

We are able by means of these and other passages to trace the custom of meeting in private houses down to the very end of the ages of persecution. In the subsequent period such gatherings were discouraged. It was against the policy of the Church to allow meetings of the faithful to be held without the superintendence of the authorities. It was believed, no doubt, that they would lead to divergencies of various kinds from orthodox dogma and practice, and they were prohibited unless licensed specially by the bishop. The institution did not however pass away without giving birth to an offshoot which has survived to modern times. This is the private oratory for family use, of which we find traces in early Christian days. On the Aventine there was discovered in the last century a small oratory, adorned with Christian paintings of the fourth century, which belonged, apparently, to the house of one Pudens Cornelianus, a member of a family famous in the early traditions of the Church. Another small oratory of interesting form, and adorned with Christian paintings, was discovered near the Thermæ of

[1] *Ante*, p. 19.

Diocletian, and is described by de Rossi in his *Bullettino* for 1876. These domestic chapels are of course quite distinct from the large private halls of which we have been speaking.

The question now arises, Were these the *only* meeting-places possessed by the Christians in the Roman cities? Eusebius tells us that at the time of the Decian persecution in the middle of the third century, there were forty-six presbyters at Rome,[1] and the statement of Optatus, that in the days of Diocletian there were more than forty Roman basilicas,[2] leads to the inference that there were that number of independent communities, each presided over by a presbyter. Did these communities only meet in private houses?

The supposition is a very natural one that, in some parts of the city at any rate, common meeting-places, unconnected with private houses, had at an early period been erected. It was not in every quarter of Rome, or in every place where there was a community of Christians, that houses with large private rooms would be available for the purposes of the faithful. In cases where these were wanting, nothing is more probable than that the Christian communities—the outward organisation of which resembled, as we have seen, that of other religious associations—should have followed the example first of the Jews and then of the various clubs and colleges which were so numerous all about them, and have procured or erected, *ex ære communi*, a modest meeting-place, such as that which we may imagine used by the Christians of Corinth.[3] In days when active persecutions were not unfrequent, it was important for these buildings to be as unpretentious as possible, and there was no way by which they could more surely escape hostile

---

[1] *Hist. Eccl.* vi. 43.
[2] *De Scism. Donat.* ii. 4.
[3] 1 Cor. xi. 17-34.

notice, than by assuming the character of the *scholæ* or lodge-rooms, already referred to as the meeting-places of the pagan associations. As a fact, the first account we have of any action of the Christians in a matter of the acquisition of land, presumably for building, suggests a *schola*. It occurs in the life of Alexander Severus (222-235) by Lampridius, and is to the following effect:—'The Christians had occupied a certain space of land which had before been public property, and the tavern-keepers (*popinarii*) opposed them, saying that the ground ought to be theirs. The emperor decided that it was better that God should be worshipped in that place in any sort of a way, than that it should be given over to tavern-keepers.'[1]

This passage should be taken in connection with another which describes the action of Severus in encouraging clubs and guilds in general. 'He established corporations everywhere, of the wine-dealers, the inn-keepers, the cobblers and, in a word, of all the trades.'[2] It is fair here to conjecture that the claim of the tavern-keepers upon a piece of formerly public land arose out of this action of Severus towards the guilds. There can be little doubt that these *popinarii* formed a society similar to those mentioned above by Lampridius, or to the '*corpus tabernariorum*,' or 'guild of the victuallers,' which we find in an inscription,[3] and it may be that Severus had gone so far in his favour towards the societies as to allow some of them to appropriate portions of vacant public land on which to build their meeting-places. Expecting a similar permission, these tavern-keepers found themselves forestalled by a community of Christians. The matter is referred to the emperor, who looks at the dispute as one between two associations, one religious, and the other trading, and decides that any sort of religious society is more

[1] Lampridius, *Alex. Sev.* c. 49.
[2] *Ibid.* c. 33.
[3] Orelli-Henzen, 7215 a.

to be encouraged than the guild of the tavern-keepers.[1] Whatever building the Christians erected on the ground thus secured would have been considered a *schola*, and would have had the form and architectural character of those buildings.

These *scholæ*, the name of which survives in the Italian term *scuola*, for the lodge of a fraternity, belong to a class of public buildings not so important as temples, theatres, baths, or basilicas, but essential to the equipment of an ancient city. Official bodies of various kinds, from the Roman or the municipal senates downwards, had their places of business or of deliberation, called *curiæ* or *scholæ*. The *curia* was the regular name for a senate-house, and the senate at Rome had its *curiæ*, though it held its meetings very commonly in temples, especially in the temple of Concord; the ancient divisions of the Roman people, too, the *curiæ*, met in edifices called by the same name. Bodies of officials, like those who managed the public games, or the details of municipal affairs, had also their proper offices, which seem to have been called *curiæ* or *scholæ*, according to their greater or less importance. Thus, a set of public clerks at Rome were located close to the Tabularium, in a row of chambers called the *Schola Xanta*, while the august priests of Mars, the Salii, had a *curia* upon the Palatine. At Ostia the semi-official guilds of officers and workmen about the port had their *scholæ*, and these have recently been brought to light—some above the magazines of grain, and others in a portico on the Forum, divided for this purpose into sundry compartments, on the thresholds of which are inscribed the names of the guilds in occupation.[2]

---

[1] This story bears upon the question, noticed in the last chapter, whether the Christians could hold a formal legal title, as a corporation, to ordinary property (p. 22, *note*).

[2] Lanciani, in the *Athenæum*, July 1880 and Jan. 1882.

Within the walls of Rome the situations of some of the meeting-places of trading or religious corporations are indicated by inscriptions, though there are no actual remains to point to. Thus we read of a college whose members met in a *schola* by the Theatre of Pompeius.[1] A religious brotherhood devoted to the worship of the genius of the Imperial family, had a *schola* above the Temple of Claudius.[2] A third was close to the Pantheon of Agrippa.[3] The fishermen seem to have been located by the Tiber bank,[4] and a writer in Grævius suggests that the *scholæ* of many of the trade guilds would have been found near the market under the Aventine.[5]

The form of these *scholæ* would doubtless vary according to the property and importance of the college. Some of them were built and adorned at the expense of wealthy individuals, whose benefactions are recorded by inscriptions, and these would possess some architectural pretension. For example, we read of a ' *schola* with its statues and images, and all its adornments;'[6] and again, ' Cornelius Aphrodistus erected at his own cost a new building as a gift to his associates, and dedicated it on the Kalends of June;'[7] ' Orfia Priscilla paid to the college of smiths the sum of money which Orfius Hermes, her grandfather, had promised for the decoration of their *schola*, in remembrance of Orfius Severus, his son.'[8]

The best idea of the form of these buildings is to be gained from the three small edifices which stand side by side close to the Forum of Pompeii, and are marked XIX. on Overbeck's plan.[9] Fig. 4 gives the form of two of them: they were simple oblong halls, measuring about 50 feet by 30,

[1] Orelli-Henzen, 4085.
[2] *Ibid.* 2389.
[3] *Ibid.* 7215 a.
[4] *Ibid.* 4115.
[5] *Thesaurus*, iv. 1537 c.
[6] Orelli-Henzen, 3303.
[7] *Ibid.* 4092.
[8] *Ibid.* 4089.
[9] Overbeck, *Pompeji, etc.*, 4 Aufl., Leipzig, 1883.

and were embellished with some taste. They probably formed the offices or board-rooms of municipal functionaries, but the *scholæ* of private associations may well have had a similar form. It may be remarked that on the fragments of the Capitoline plan of Rome, there are not a few buildings indicated which have the same form as those of Pompeii, and may have been *curiæ* or *scholæ*. Fig. 5 [1] gives some specimens. It will be observed that a constant feature in these buildings is the apse at the small end furthest from the entrance, which was without doubt the place set apart

Fig. 4.—Curiæ or Scholæ at Pompeii, from Overbeck.

for the president and officials at the meetings; and the similarity of this arrangement to that prevailing in the Jewish synagogue, and at a later time in the Christian Church, will not escape attention. A good idea of the general appearance of small buildings of this kind may be gained from Fig. 6, taken from an early Christian mosaic, which possibly (*infra*, p. 99) represents a Jewish meeting-house. It is conceivable that some of the halls shown in

[1] Jordan, *Forma Urbis*, Frag. 228, 233, 236, 289, 386.

plan in Fig. 5, may have sheltered Jewish or Christian communities.

The brotherhoods whose place of meeting was outside the city, were generally funeral societies, and their *scholæ* were connected with the cemeteries. Outside the walls of Rome more than one such ancient *schola* has been identified. The foundations of two have been discovered among the tombs on the Appian Way, and there is no doubt that they served for the burial and memorial feasts of the funeral colleges.

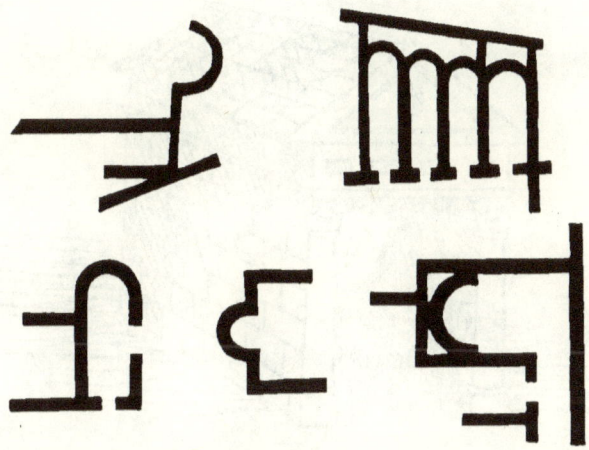

FIG. 5.—Scholæ (?) from the Capitoline plan of Rome.

One of these is the *schola* mentioned in an inscription quoted in the last chapter (p. 18) as belonging to the college of Silvanus. It lay on the right of the Appian Way, a little beyond the cemetery of S. Callisto, and was of a circular form, with an altar in the centre, and seats of stone ranged round the circle of the walls. A portico was also connected with it.[1] A second *schola* is described as a quadrangular *cella* with a single entrance, measuring about 18 feet on every side. In the midst was an altar, and around the

[1] Carlo Fea, *Varietà di Notizie*, p. 180.

walls ran a continuous bench of stone, which would have accommodated nearly fifty persons.[1] Fig. 7 gives the plan of the remains of another *schola* belonging to a pagan religious brotherhood, which was discovered at Ostia.[2] Here we have an irregular quadrilateral plan with two altars in the midst, and a continuous bench of the same kind as before. In the case of the Christians, on the other hand, de Rossi has been fortunate enough to discover and identify the remains of a large *schola*, very similar in form

FIG. 6.— Small Synagogue or Schola, from a Mosaic of the Fifth Century.

and arrangement to the last, by the entrance to the ancient catacomb of Domitilla. It was of irregular form, and possessed a stone bench along the walls; a vaulted chamber with a well opened out of it. This de Rossi calls 'a vast triclinium for a large number of guests, in a word, a *schola sodalium* similar to those of the pagan brotherhoods instituted for purposes of burial.'[3] The parallel between Christian and Pagan could not be more exact.

[1] C. L. Visconti in *Annali dell' Inst. di Corr. Arch.*, 1868, p. 387.
[2] *Annali del Inst. Arch.*, 1868, p. 362.
[3] De Rossi, *Bullettino*, 1865, p. 97.

PAGAN AND CHRISTIAN. 55

Akin to the *scholæ* of these burial associations, were the *cellæ* erected in cemeteries for the due performance of funeral rites. Directions for the fitting up of one have already been given in the words of a pagan testament.[1] Here again Christian sources furnish us with something closely parallel, and we come now upon the traces of an important form of early Christian building. A Christian inscription from Cherchel in North Africa, quoted on a previous page, records the presentation to the church of a

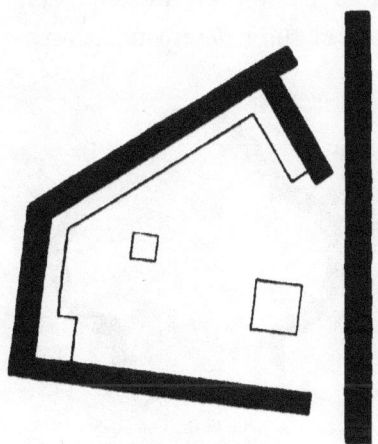

FIG. 7.—Schola beside the Metroon at Ostia.

place of burial with a *cella*.[2] Eusebius[3] quotes a writer of the end of the second century who tells us that in his day the 'trophies' (memorial chapels) of the martyrs Peter and Paul were to be seen on the Vatican Hill, and on the road to Ostia. The *Liber Pontificalis* states that Pope Fabian (236-250) 'caused many structures to be erected in the cemeteries,'[4] and it is probable that most of these were memorial *cellæ*. The researches of de Rossi in the region of the catacomb of S. Callisto at Rome have identified two of

---

[1] *Ante*, p. 19.   [2] *Ante*, p. 25.   [3] *Hist. Eccl.* ii. 25.
[4] Anastasius, *Hist. de Vitis Rom. Pont.* 21.

the ancient buildings, which remain above ground at that spot, as Christian *cellæ* of the kind referred to in the inscription from Cherchel.[1] These seem to date in their original form from the period before the Diocletian persecution, and may contain some of the original work of Pope Fabian. The form of these edifices is of great architectural interest. It will be convenient to deal with the structural questions connected with them in the next chapter, and to aim only in this place at bringing their probable aspect and use as clearly as possible before the reader. Fig. 8 shows the ground plan of one of the *cellæ* upon the area of the cemetery

FIG. 8.—Plan of Memorial Cella in the Cemetery of S. Callisto, Rome.

of S. Callisto, in the original form as ascertained by de Rossi, while the drawing upon the frontispiece of this volume is an attempted restoration of a similar building, as it may have appeared at the time of the preparation for a memorial feast in honour of a martyr in the latter half of the third century A.D. The ground plan shows three semicircular apses set trefoil-fashion on the three sides of a square, the fourth side being left open. Such a structure would be aptly described by the term *exedra* (alcove), a word used in the Roman testament (p. 19) to express the place appointed for the memorial feast on the birth-

---

[1] De Rossi, *Rom. Sott.* iii. 468.

day of the testator, so that the arrangements described in that curious document may be taken as data to assist in the reconstruction of the memorial *cella* of the Christians.

In the drawing on the frontispiece, the *cella* is supposed to be situated in a Christian cemetery on rising ground above the Appian Way, a short distance beyond the gates of Rome, the buildings of which are seen in the middle distance. The area of the cemetery is bounded by walls, with gates for access on the two sides nearest the city, and is pleasantly laid out as a garden.[1] Within it are to be observed, (1) a memorial *cella*, (2) an entrance to the subterranean galleries and chambers, which were the actual places of sepulture, and (3) the small tenement where the officials of the cemetery might reside, and where would be kept all the furniture and utensils required for the funeral rites and the memorial feasts. These officials formed an important body, called *Fossores*, who were numbered among the inferior clergy,[2] and charged with the excavation of tombs, and the care of the cemeteries in general. Portraits of these *fossores*, bearing their picks for excavation and their lamps, occur in the wall-paintings of the catacombs, and in the case of one interesting picture, which seems to portray the celebration of the Eucharist and the *agape*, a figure of a *fossor* occurs at each side, standing with pick on shoulder, as if on guard to prevent the intrusion of the uninitiated.[3] Hence we may suppose that the *fossores* of a cemetery would not only carry out the arrangements for a memorial feast under the direction of the deacons, but would also, as shown in the drawing, guard the gates while the congregation was assembling.

---

[1] The arrangement of the Christian cemeteries *above ground* is discussed by de Rossi, *Rom. Sott.* iii. 400 ff.

[2] As in the description quoted on page 46.

[3] Kraus, *Real-Encyklopädie*, art. 'Eucharistie,' p. 441.

With regard to the *cella* itself, its dimensions were small, that shown in plan in Fig. 8 measuring about 30 feet on each side of the square. The large congregations that assembled on the festivals of the martyrs must have found accommodation on the open space outside, while the religious service and the preparation of the feast were directed by the presbyters and deacons within the building. Such an open-air feast has classical parallels. For example, in the street

FIG. 9.—Funeral Triclinium at Pompeii, from Overbeck.

of tombs at Pompeii, there is a small open *triclinium* arranged with couches for the guests at a funeral banquet. Fig. 9, from Overbeck's *Pompeji*, gives a view of this, and the arrangement of the three sloping couches, with the table in the centre, has been adopted in the restoration on the frontispiece. Couches and tables might be, like those at Pompeii, of brick or stone covered with stucco, or the couches might be of turf, with moveable tables of wood. In the drawing, couches and a table for the clergy and distinguished members of the congregation are placed in front of the *cella*, while seats, couches and tables for the main body of the people stand at the sides. A division formed of wooden barriers runs down the centre to keep apart the women and the men.

For the furnishing forth of the memorial feast there would be required the cushions, rugs, etc., mentioned in the Roman testament, as well as the wine-jars and other vessels included in the inventory of the possessions of the Christian community at Cirta, quoted on p. 46. These would be under the charge of the *fossores* who kept the cemetery. With respect to the *cella* itself, as shown in the drawing, its ground plan is, as we have seen, derived from an actually existing ruin. The façade is copied from that of an *exedra* in the street of the tombs at Pompeii.[1] The exterior is adorned with inlaid marbles, the façade with reliefs representing the story of Jonah, the interior with wall paintings.[2]

---

[1] Nicolini, *Case e Monumenti di Pomp. descriz. gen.* Tav. vi.

[2] The best early patristic authority for these memorial feasts with which the writer is acquainted is to be found in the so-called Canons of S. Hippolytus, edited by de Haneberg from an Arabic MS. in the Vatican Library, under the title: *Canones S. Hippolyti Arabice e Codicibus Romanis cum versione Latina, etc.* Monachii, 1870. Canon thirty-three runs as follows:—'*About the commemoration for the dead; it is forbidden to hold it on the Lord's day.* If commemorations are held for those who are dead, first, before the people sit down to meat, let them take the sacrament, but not however on the Lord's day. After the communion, let the bread (of thanksgiving) be distributed to them before they sit down to meat, but let it be done before the sun sets. Let no catechumen sit down with them in the Agape. Let them eat and drink as much as they need, but not to drunkenness, but as in the divine presence, and with thanks to God.' The Latin version is appended:—

'XXXIII. *Canon trigesimus tertius. De commemoratione pro defunctis: interdicitur, ne fiant die dominica.*

'Si fiunt commemorationes pro iis, qui defuncti sunt, primum, antequam consideant, sacramenta sumant, neque tamen die dominica. Post communionem distribuatur eis panis (eulogiarum) ante solis occasum,

Next to the *scholæ* of the cities, and the memorial *cellæ* on the cemeteries, must be mentioned the underground chapels excavated in the *areæ* of the burying grounds, and used, as we have seen, as places of meeting during the dark hours of the Church's history in the Decian and Diocletian persecutions.[1] The form may be gathered from the accompanying illustration (Fig. 10) representing one discovered in the catacomb of S. Agnese, and described and figured in Marchi's work on the early Christian cemeteries. The space is formed by throwing into one a number of the small square cells or

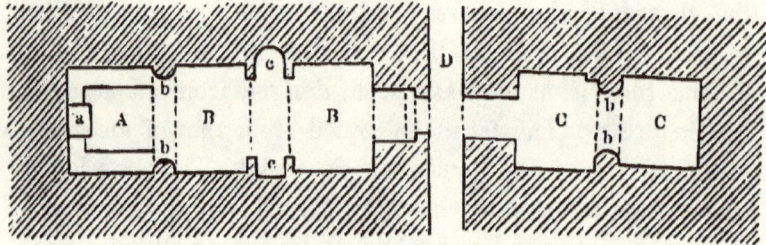

Fig 10.—Underground Meeting-place in a Roman Cemetery.

antequam consideant. Non sedeat cum eis aliquis catechumenus in Agapis.

'Edant bibantque ad satietatem, neque vero ad ebrietatem; sed in divina præsentia cum laude Dei.'

It is clear from this passage that the memorial feast in honour of the martyrs, called here by the name *agape*, was a recognised institution, and was held without any secrecy; the author of the Canons is particular in his directions that all is to be done by daylight, so that the people may return safely and in order to the city before nightfall. The double character of the celebration (which begins with the religious ordinance of the Eucharist, but tends as we see from the last sentence to become over-convivial), and the exclusiveness of the gathering (which includes only those in full church-fellowship), are also points to be noticed.

The above Canons may be ascribed to the first half of the third century (de Haneberg, p. 22). At a later period these *agapæ* are discouraged (S. Greg. Naz., Migne, S. G. xxxviii. p. 98; S. Chrys. *in S. Jul. Mart.* § 4), and the councils of Laodicæa (364 ?), (Labbé, ii. 570) and Hippo (393), (Labbé, iii. 923) decree against them.

[1] The Chronological Table gives a view of these successive periods of repose and persecution.

cubicula, in the walls of which were constructed the niches to contain the bones of the honoured dead. The cubiculum A is the seat of the clergy, with the bishop's chair at a. B, B, are the spaces for the men; C, C, those for the women of the congregation. The gallery D gives access from the other parts of the cemetery. The pavement of the church, if so we may call it, was of marble, and pilasters projecting from the wall at b, b, b, b show some attempt to give architectural character to the construction. The niches, c, c, are supposed by Father Marchi to have held statues, but were more probably receptacles for lamps. If this archæologist is right in his explanation of this crypt, and in assigning to it a date in the third century, we have here for the first time some of the peculiarities of arrangement which we find in force in the churches of the age of Constantine. They concern the separation of the sexes during worship, an arrangement insisted on by the Church, and one for which the Christian, like the Jewish, architect had to provide. St. Chrysostom tells us that of old such separation was not necessary. In Christ there is neither male nor female; and in the upper chamber at Jerusalem there were gathered together both men and women. Now, however, he explains, some purity of heart has been lost, and it has been found needful to place wooden barriers in the churches to keep the sexes apart.[1]

Another feature is the division of the presbytery, or place for the clergy, from that appointed for the people. The former were ranged on stone benches on each side of the bishop, who probably had in front of him a portable altar. There is nothing which has had more influence on the planning of Christian churches than the necessity of providing for this separation, and it is interesting to observe it here marked for the first time in existing monuments.

[1] *Homilia* lxxiii. *in S. Matthæum*, 3.

That the Christians were driven in times of persecution to meet in still less convenient localities than these sepulchral crypts will readily be believed. 'Dens and caves of the earth' were put to sacred uses, and Gregory of Tours tells us that the few followers of the first bishop of that district met in crypts and caverns in the time of the Emperor Decius.[1] The *speluncæ* (disused cisterns ?) of Alexandria sheltered the trembling congregations in the days of Aurelian.[2] These temporary expedients have no importance for our present purpose, and the synagogue, the private hall, the schola, the memorial cella and the crypt, remain the only forms with which we need concern ourselves.

The circumstances under which the formal church (whatever may have been its architectural form) came to supersede these irregular structures must now be considered. The dedication of private halls as Christian churches has already been noticed. Only in a single instance has such a hall been preserved till modern times. Up to the period of the Renaissance there existed at Rome, in the vicinity of S. Maria Maggiore, an ancient church known as S. Andrea in Barbara, remarkable for its splendid internal decoration in mosaics of coloured marble. The researches of de Rossi[3] leave little doubt that this was originally erected in 317 A.D. by a wealthy Roman, the consul Junius Bassus. For what purpose it was built we cannot exactly determine, but the subjects of the mosaics, which were taken from pagan mythology and history, leave no doubt that it was once a heathen secular building. An inscription in the apse proves that in the latter part of the fifth century it was the possession of a private person of wealth, who bequeathed it by testament to the Church. The inscription describes its

---

[1] Greg. Turon. *Hist. Franc.* x. 31.
[2] Eutych. Alex. *Ann.* i. p. 398 ; Migne, S. G. cxi. 997.
[3] De Rossi, *Bullettino*, 1871, p. 5.

dedication to Christian worship by Pope Simplicius (468-483), who was content to substitute a fine Christian mosaic in glass for the old decoration of the upper part of the apse, and to leave the rest in its original state. The loss of this building, which had fallen into ruin before the seventeenth century, deprives us of a unique and most interesting monument of early Christian times.

In the case of not a few of the principal Roman churches we have traditional or other evidence to connect them in their origin with private houses. In these instances what actually remains to us is a church in the regular form of the fourth or fifth century, no portion of which seems ever to have done service as a private hall; and we may suppose, if this derivation be correct, that the mansion would in time fall into ruin and disappear, while the hall would be altered and rebuilt to suit ecclesiastical requirements. Thus the famous Basilica of St. John Lateran, 'the mother church of Christendom,' was in its earliest form a portion of the palace of Fausta, the wife of Constantine; and an apartment of the Sessorian Palace became the Church of Santa Croce in Gerusalemme. According to tradition the ancient Church of S. Pudentiana, one of the oldest in Rome, stands on the site of the palace of the senator Pudens, the supposed entertainer of St. Peter. A legend connects S. Cecilia in Trastevere with the house of the saint herself, and in the case of S. Clemente we have the interesting fact that the lower church, which dates from the days of Constantine, is connected with the chambers of a private mansion, now buried beneath the ground, which is conjectured to have been the abode or the place of teaching of St. Clement of Rome. It is true that these are somewhat shadowy personages, but the traditions are not so valueless as to warrant our omitting all notice of them.

For the derivation of Christian churches from synagogues,

or from the ordinary city *scholæ*, we have no direct evidence, though a tradition ascribes to the Church of S. Maria in Trastevere an origin in the building which was the subject of dispute between the Christians and the *popinarii* in the days of Alexander Severus. In the matter of the buildings in the cemeteries the evidence is much clearer. We have seen that meetings under ground were only rendered compulsory through exceptional edicts of persecution, and the subterranean churches were not used more than was absolutely necessary. Though they are in themselves interesting memorials, they are too small and have too little architectural character to have contributed much to the ultimate forms of the Christian church. The contrary is the case with the memorial *cellæ* above ground, the connection of which with later architectural efforts can clearly be traced.

It will readily be remembered that many of the principal Roman basilicas are situated outside the walls of the ancient city. This was the case with the two mightiest of them all, St. Peter's on the Vatican and St. Paul's on the road to Ostia, as well as with the ancient shrines of S. Lorenzo and S. Agnese. These churches originated neither in private houses nor in lodge-rooms, but in memorial *cellæ* of the martyrs whose bones were supposed to rest beneath them. The enthusiasm of the age brought ever-increasing throngs of the faithful to celebrate their feasts, and structures, which at first were only *exedræ* or chapels, had grown by the fifth century into enormous churches. The process of this growth may be seen in a curious monument a few miles from Rome. On the road to Tibur, nine miles from the city gates, lay the spot hallowed by the bones of the martyrs S. Sinforosa and her seven sons. Their remains were laid in a tomb, above which was built an *exedra* or *cella* (A. Fig. 11), with three apses of the same form as those which still exist on the area of S. Callisto. In course of time, this accommodation proving insufficient, a regular church of the basilican form was erected close beside it at B. The importance of proximity to the martyrs' grave was so great that the buildings were placed in contact with each other back to back, and an opening was made through the walls at *a*, so that worshippers in the apse of the basilica might still be able to look down upon the spot where the sacred bodies were laid.[1] The arrangement is an instructive one; it is an early manifestation of a feeling which had large influence on later architecture, —the feeling that a church and an altar are hallowed by contact with the body or relic of a saint.[2] The fantastic tricks which were played with relics even in early days, and continued through the Middle Ages, have their origin in this reverence for even the bones of the martyrs. Churches erected in the place of the original memorial *cellæ* derived their importance from the fact that they guarded the relics of the martyr, and there are numerous instances of the extreme care taken by architects in the reconstruction and enlargement of churches, to avoid changing the situation of the altar in its relation

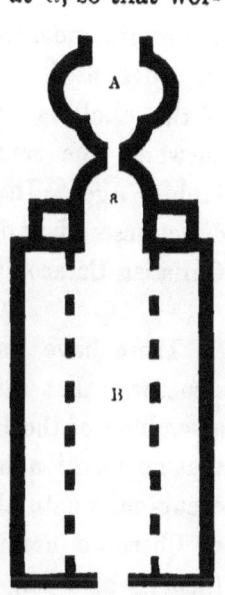

Fig. 11.—Cella and basilica of S. Sinforosa, near Rome, from Stevenson.

---

[1] H. Stevenson, in de Rossi's *Bullettino*, 1878, p. 75.

[2] A precedent for this outside the Christian pale can be found in the customs of the Synagogue. A dead body would have been considered to defile the Jewish Temple, and the Greeks and Romans kept their sepulchres apart from their towns and their sacred enclosures. Funeral ceremonies involving the presence of the corpse were, however, performed in the synagogues (*Jerusalem Berakhoth*, iii. 1, Schwab, i. 57), and it became the custom for these buildings to be connected with the tombs of their founders or of some great man of the past. Interments did not take place in the synagogues, but beside them, where the tomb remained to add a dignity to the place of meeting.

to the sacred tomb. When the actual visits to the burial-places without the walls fell into disuse there ensued a curious change. The Church, no longer able to go out to honour the martyrs, brought the martyrs in to herself within the walls, and instead of building churches above the tombs, dug tombs under the churches in which the precious treasures were deposited. This was the origin, first of the *confessio* of the basilicas, and at a later period of the crypt which answered the same purpose in the Church of the early Middle Ages. In this way the Romanesque crypt is the direct descendant of the *hypogæum* or excavation of the early Christian Catacomb.

There have now been passed in review the principal structures that can be brought into connection with the assemblies of the Early Christian Church. It will be noticed that no mention has yet been made of the building which occurs most naturally when a question arises about the origin of Christian architecture—the pagan basilica. The old theory which derives the Christian Church from the basilica will be noticed more at length on a subsequent page. Its insufficiency is seen at once when we consider that it was not till the Church had grown comparatively wealthy and strong that buildings of the size and importance of basilicas could ever have been thought of, and as we cannot suppose the history of Christian architecture for the first two hundred years to have been entirely a blank, we must look to other buildings than basilicas for its first beginnings. Now no one of the structures just noticed can, any more than the pagan basilica, be taken by itself as a model for the Christian temple. Christian architecture had, as it were, several root-fibres traceable back to non-Christian origins, and these unite to form the historical Church of the age after Constantine, an architectural creation embodying features derived from different sources. There is, to begin with, the Synagogue. Remembered as it was with affection by all Jewish converts, and designed to suit a congregational service resembling in many important points that of the Christians, what could seem more suitable as a model for the meeting-place of the brethren ? The primitive Christians, assembling when and where they best could, and exposed from time to time to actual molestation, might be excused if they turned with a sigh of envy to the secure and well-appointed synagogues of the older faith, and it is perhaps with a special view to their encouragement that the author of the Apocalypse describes the New Jerusalem as dispensing with any visible shrine of prayer.[1] As the Christians grew strong a feeling of rivalry with the Synagogue may well have influenced the constructors of their churches, and led to a certain assimilation of their form to that of the older edifice. On the other hand, the strong antagonism between Christians and Jews and the Gentile origin of most of the former would tend to minimise this influence of the Synagogue, and it would probably be a mistake to attach much weight to it. The question is rendered more difficult by the fact that, as will be seen in the next chapter, we know comparatively little about the normal form of the ancient synagogues of the Jews.

The private hall is of far greater importance to our subject, for here we know the Christians habitually assembled, and we shall see reason for believing that some of the characteristics of the perfected Christian Church may be best explained as a reminiscence of the primitive meetings in dwelling-houses. The halls themselves had no one fixed shape, and none of their forms save one would, so far as we know, have provided the special features which have become stereotyped in Christian architecture. This exceptional

[1] Rev. xxi. 22.

form is that of the so-called *private basilica*, or hall of audience, of the Roman princes and nobles, and on this a word will be said later on. The building which, beyond all the others, we should be inclined to fix upon as the model of the Christian Church of the ages of persecution, out of which grew the great basilicas, is the *schola* or lodge for the meeting of associates. This had the advantage over the private hall that it was specially designed for the purposes of assembly, and its internal arrangement was one which would lend itself easily to Christian worship; the Christian would take the place of the pagan altar, and the elders or overseers of the church would occupy the same position of presidency as the corresponding officials of the *sodalicia*. Hence a *schola* would give the Christians exactly what they needed, and would only need enlarging in the manner to be studied in the next chapter to become the Church of the fourth century. It is unfortunate that this can only be presented as a conjecture. A city *schola* of the Christians has yet to be discovered, but it is hoped that the hypothesis that *the city schola was the earliest form of building erected by the Christians with a special view to their gatherings* has sufficient intrinsic probability to admit of its being put forward in this place.

The private hall was not *erected* but only *used* for Christian worship, the *schola* (supposing the above hypothesis to be correct) would be designed for the purpose, but would be a simple building with *use* for its primary object, and would not be in any sense the architectural expression of Christian ideas. For such expression we must turn elsewhere, and we find that a new and important element in the development of Christian architecture is introduced by the *exedra* or memorial cella of the cemeteries, for with these that architecture takes for the first time a pronounced monumental character. The buildings we have been discussing, whether expressly erected or only adapted for worship, were to the Christians merely buildings for accommodation; that is to say, there was no connection between the feeling of the Christian community and the architectural features of its buildings. To the worshippers these were mere accidents; all that they required was a meeting-place providing them with the needful shelter and privacy, and arranged on a convenient plan. In the memorial cella of the cemeteries, on the other hand, we have at once architecture of an expressive kind, a building erected not so much for *use* as for *feeling*, for the celebration, that is to say, of an idea which the community had at heart, the commemoration of its mighty dead. The name applied to these cellæ in a passage already referred to[1] is significant. They are called in Eusebius 'Trophies of the Martyrs,'[2] so that their erection as a heroic honour to the martyrs made them *the first definite architectural expression of Christian feeling*, and as such their influence will be found very apparent in the perfected form of the Christian temple.

Some answer has now been gained to the questions, In what buildings did the primitive Christians assemble? How did these develop into the regular church? and, Which of these buildings is most likely to have given that church its special form? and there only now remains to consider, At what period did the regular church first become an established institution? The 'Age of Constantine' is a convenient era, and is generally taken to be the period in question; but we should probably be more correct in antedating it by some half-century. Considerable light is thrown upon the condition of Christian architecture in the half-century before the age of Constantine by a passage in

---

[1] *Ante*, p. 55.
[2] The term *trophies* in this passage must certainly mean some conspicuous monument above ground, an *exedra* or *cella*.

Eusebius referring to the forty years' peace of the Church, which preceded the breaking out of the Diocletian persecution in A.D. 303. We there read of the 'innumerable multitudes of men who flocked daily to the religion of Christ,' and of the 'illustrious concourse of people in the sacred edifices,' whence it came about that, 'no longer content with the ancient buildings, they erected spacious churches from the foundation in all the cities.'[1]

If the 'ancient buildings,' now discarded, were the private halls or the *scholæ* already described, the 'spacious churches' must have possessed a distinct and imposing form. A saying put into the mouth of the Emperor Aurelian (270-275) proves that in his time they were well-known features in Roman cities. 'It would seem,' said the Emperor reproachfully to the Roman Senate which had displeased him, 'that you were holding your meetings in a church of the Christians instead of in a temple of all the gods.'[2] For a reason which will presently appear, we have no specimens remaining of the church architecture of this particular period, but it may be conjectured that it would approximate more or less closely to that we find established in the days of Constantine. At one of these buildings we are accorded, however, a passing glimpse in an account of the opening scene of the great Diocletian persecution, when edicts were put forth to 'tear down the sacred buildings to the foundation, and burn the Holy Scriptures with fire.'[3] Lactantius records the destruction on this occasion of the important Christian church at Nicomedeia, the residence at the time of Diocletian, and the seat of empire. It was situated on rising ground,[4]

---

[1] *Hist. Eccl.* viii. 1.   [2] Vopiscus, c. 20.
[3] Eusebius, *Hist. Eccl.* viii. 2.
[4] We are reminded here of the Jewish synagogue, which, according to the Talmud, ought to occupy the most elevated position in a town or village.

within full view of the palace, and was surrounded by many and large buildings. The gates were first forced open and the building pillaged, after which Diocletian, dreading a general conflagration if it were set on fire, lets loose upon it the Pretorian guards, who, with axes and crowbars, in a few hours 'levelled that very lofty edifice with the ground.'[1] Partly owing to the general carrying out of these edicts of destruction, and partly to the rebuilding which went on in the succeeding age, it is impossible to point to any existing church of an earlier date than the time of Constantine, and it is not till then that the monumental history of Christian architecture has its beginning.

The reign of Constantine introduced an entirely new era for the Church. Christianity was now for the first time placed openly, and in all its relations, under the protection of the law, and the important right of receiving legacies accorded to it. Whatever was the personal attitude of Constantine towards Christianity, it fell in with his policy to interest himself in ecclesiastical affairs, and give them all the prestige of Imperial favour. The Church took at once full advantage of its changed position, and proceeded to surround itself with the utmost possible pomp and splendour. The meeting-place at Nicomedeia, though lofty and spacious, can only have been a slight structure, if a few hours sufficed for its demolition. Constantine, if we may accept the accounts of Church historians, now set the example of erecting solid structures of squared stones, and extending to vast dimensions the houses for Christian worship. The faithful threw themselves eagerly into the work, and, as Eusebius writes: 'Every place that a short while before had been desolated by the impieties of the tyrants, revived again, and temples rose once more from the soil to a lofty height, and received a splendour far exceeding that which had been

[1] Lactantius, *De Mortibus Persec.*, 12.

formerly destroyed,' while 'there were festivals and consecrations of newly erected houses of prayer throughout all the cities.'[1]

This outburst of zeal for building coincided with a revival in the arts, which enabled it to find a fitting outward expression. Among the edicts of Constantine we find some which relate to measures for the encouragement of architects and decorative artists, whose services the Emperor wished to employ on his vast undertakings at the new capital of Constantinople. 'Of architects,' he says in one of his latest edicts, 'there is great need.'[2] To the new Rome he transported a large number of statues and other works of art from the cities and temples of Greece, and through the influence of these models, as well as through the measures taken for the training of artists, the arts, which had been rapidly declining, received a new impetus. The successors of Constantine continued to encourage the artists, and reaped the fruit of the movement which he had set on foot. Some important edicts of Theodosius are evidence of the interest taken by that great emperor in artistic matters.

The Church was not slow to avail itself of the opportunities thus brought within its reach. It is absurd to suppose that Christianity would in itself have brought about a revival of art. Without good models and good training, Christian artists would have produced a barbarous, though perhaps expressive art, like that represented at a later time by the bronze doors at Hildesheim. The classical revival between the time of Constantine and that of Theodosius kindled anew the old traditions of Roman building, and supplied the designer with a conception of the human figure and of drapery that had something of antique simplicity and

---

[1] *Hist. Eccl.* x. 2, 3.
[2] Const. Mag. *decreta*, ed. Migne, p. 381.

grandeur. With the stimulus upon him of the new impulse, and at the same time with the command of all ancient lore, the Christian architect girded himself to his task, and created forms which were destined to have the most splendid history in architectural annals, while the decorative artist, following in his steps, elaborated a style of graphic design as expressive and beautiful as it was original. The modes of architectural and artistic expression which were in this way adopted, it will now be our task to trace.

CHAPTER III.

THE BASILICA, THE SYNAGOGUE, AND THE CHRISTIAN CHURCH.

ON the carved sarcophagi, and in the mosaic pictures of the period after Constantine, occur representations of the Christian churches of that day, of which Figs. 12, 13, 14 may be taken as specimens. Fig. 12 shows a group of ecclesiastical buildings. In the foreground is a hall of simple plan, suggesting a *schola*. Further back is a similar building, terminating in an apse, like the *curiæ*, or kindred edifices, at

FIG. 12.—Early Christian buildings, from a carved sarcophagus. Fourth Century.

Pompeii, or on the Capitoline plan of Rome. By the side of it is a small round structure covered with a domed roof, and a similar pair of buildings is seen in the right-hand corner. The oblong building is a meeting-house, the other a baptistery, wherein was administered the rite of initiation to the Christian fraternity, and the circular form of which marks a new departure in Christian architecture. Two buildings of

EARLY PICTURES OF CHRISTIAN CHURCHES. 75

the same kind appear in Fig. 14, from a sixth-century mosaic in S. Apollinare Nuovo at Ravenna, but there is here an important architectural difference. The groun plans, it will be observed, are no longer simple, but side-aisles or their equivalent are added, with the effect of

FIG. 13.—Ecclesiastical structures, from a fifth-century mosaic in S. Maria Maggiore at Rome.

largely increasing the available area. These side-aisles, or in the case of the baptistery this circular surrounding passage, must be understood to be divided off from the central space by colonnades. Above these rise the walls of the central portion, which, as they ascend above the side-roofs,

FIG. 14.—Basilica and Baptistery at Ravenna. Sixth century.

are pierced with large clerestory windows for lighting the interior. This central portion, and the side spaces, have distinct roofs, which are covered with tiling. The oblong building has a porch at the entrance end, and at the other the familiar apse. In this interesting little picture, reproduced here from an accurate drawing by Mothes,[1] we have a contemporary representation of the early Christian church in the perfected form which it had assumed by the days of Constantine, a form retained, though with many modifications, through the whole history of ecclesiastical architecture, and the first adoption of this special type of construction marks the moment when the Christian church takes its place among the distinct architectural types of the world.

We have already seen that the adoption of this type of building took place in all probability during the latter half of the third century, when the enormous influx of converts necessitated a corresponding increase in the size of the meeting-houses. The source from which was derived this simple but ingenious plan of side-aisles and clerestory lighting must now be considered.

This source may be briefly indicated as the so-called BASILICA. It was from the buildings known in classical times as Basilicas that appear to have been derived the distinguishing features which mark off the structures of Figs. 13 and 14 from those of simpler type in Fig. 12, and in the earlier illustrations in this book. The Basilica was, however, a building of such importance, both in itself and in its relation to Christian architecture, that it must be treated of with some fulness.

What was a Basilica?

A Basilica was a pagan secular building used like a modern exchange for the transaction of business, and containing also accommodation for the holding of courts of law.

[1] *Die Baukunst des Mittelalters in Italien*, p. 192. Mothes considers it likely that the round building is intended for the existing Catholic baptistery, while the meeting-house represents the famous 'Basilica Ursiana,' the original structure replaced by the present Cathedral of Ravenna. The Catholic baptistery is, however, of a different form from that shown in the mosaic.

A special architectural character seems to have belonged to it, and to have consisted mainly in the division into nave and aisles, and the clerestory lighting already noticed, so that all buildings with these peculiarities would be naturally called by the name *basilica*.

Such at least is the inference which it seems just to draw from sundry passages in ancient authors. Vitruvius tells us that among the different kinds of private halls, *oeci*, in the mansions of the wealthy, was one called 'Egyptian,' which was flanked upon the two long sides by colonnades supporting galleries open to the outer air. Along the interior edge of these galleries was a second row of columns, supporting a roof over the central space, and affording free passage into the interior for light and air. To make his meaning clear, Vitruvius states, in conclusion, that such structures have more resemblance to basilicas than to other private halls.[1] Again, we read that warehouses for wine[2] were sometimes called basilicas, when there could have been nothing to justify the term except special features of construction. The magnificent range of triple colonnades, built by Herod along the southern edge of the Temple platform at Jerusalem, was called by the same name, and we know that this elevation of the central portion above the two side aisles was a feature of the structure.[3] An inscription found in Britain, and dating from about 220 A.D., mentions the erection of a basilica for the purpose of exercising cavalry,[4] by which we must understand a large riding-school or drill-hall constructed on the basilican plan.

Both the name and the form of the pagan basilica have

---

[1] *De Arch.* vi. 3. 9.   [2] Rutilius, *de re rustica*, i. 18.
[3] Josephus, *Ant.* xv. 11. 5.
[4] Orelli-Henzen, 6736. '... *basilicam equestrem exercitatoriam* ... *ædificavit.*'

given rise to no small controversy.[1] The name is Greek, and means 'royal.' Some modern critics are disposed to see no special significance in the term, and take it to imply simply 'splendid' or 'costly.' Others have held that the name was borrowed from that of a conspicuous public building at Athens, the hall of the king Archon, or *Archon Basileus*.

Now there is a great inherent probability of a connection between this Athenian columned hall and the Roman basilica; and when we consider the circumstances under which the basilica was introduced at Rome, there can be little doubt that the Athenian building was its true origin.[2] The first basilica was erected at Rome by Cato the Censor, in 184 B.C., just at the period when Greek fashions began to influence the uncultured dwellers on the Seven Hills. Cato, though a Roman of the Romans, had something to do with the introduction of a taste for things Hellenic. He had served in Greece, and resided for some time at Athens. When he left he made a speech to the people, expressing his delight in the beauty and grandeur of their city.[3] Later in his life he held the censorship at Rome, an office which gave him much to do with public works, in which he could put in practice the lessons he had learnt at the metropolis of the arts. Amongst his buildings was a basilica[4]—the first

---

[1] Zestermann, in *die antiken und die christlichen Basiliken*, Leipzig, 1847, began the controversy, which has chiefly been carried on by German scholars. The Appendix to this book contains a notice of this controversy, and an examination of some recent contributions thereto.

[2] It is not likely that the Romans of the Republic would have used the term *basilica*, 'royal,' as a mere epithet with the meaning 'splendid.' Plautus, it is true, employs the adjective, but only in an ironical sense, not as implying real admiration. The use of the word in Latin demands that a more definite meaning be attached to it, and the view here maintained is, that the Roman basilica was called after the Athenian prototype, while subsequently the name was applied to other buildings on the ground of structural resemblance.

[3] Plutarch, *Cato Major*, 12.   [4] *Ib. l.c.* 19.

which had been erected at Rome—and there can be little doubt that both the name and the general form of it were borrowed from the Athenian edifice. What was the exact form, however, either of the Athenian στοὰ βασίλειος, or of this Basilica Porcia built by Cato, we have no certain knowledge, for there are neither existing remains nor contemporary descriptions on which to base an argument. It will be necessary therefore to treat the subject in a somewhat broad manner, in connection with the general character of the buildings of this epoch.

The origin and architectural affinities of the basilica will be best understood if we consider it, not as a specially Roman building, but more generally as the perfected form of a type of structure common throughout the Greco-Roman world. In other words, we must endeavour to fix the place of the basilica in ancient architecture, and by so doing, to bring into view some of the links by which the Christian temple is connected with preceding architectural structures. As this same process will be required when we pass on to consider the other form of the Christian church—its circular and domed form (Fig. 14)—it may avoid repetition if there are introduced in this place a few general remarks upon some of the aspects of ancient architecture which bear most closely upon Christian building.

The ancient world was familiar with interiors of two kinds, those connected with columned structures, and those formed by the employment of vaulting. The difference between the columnar and vaulted styles is one of great architectural importance, and is, indeed, the fundamental fact in the history of construction.[1] To build with columns is, where proper materials are at hand, a comparatively simple matter;—upright supports, joined together by hori-

[1] It is used as such by Gottfried Semper in his famous work, *Der Stil*, 2 Aufl., München, 1869.

zontal beams, all of which may be either of wood or stone, with a roof of stone slabs or timber above, are all that this style requires. The substitution for the flat or timber roof of a vault of some kind of masonry changes at once the character of the structure. Vertical support no longer suffices, and lateral pressure, due to the outward thrust of the vault, has to be provided for. Columns can offer but slight resistance to pressure from the side, and this must be met by massive walls, or their equivalent in ingeniously contrived buttresses. Constructive problems at once present themselves for solution, and unless a vast bulk of material be employed, nice calculation is necessary in order to afford support to the vault at exactly the proper point.

These two contrasted styles are represented in the works of the two most ancient peoples of which the student of western architecture need take account—the Egyptians and the dwellers in Mesopotamia. In view of the huge unbroken masses of pyramid and pylon, we can hardly describe the Egyptian as essentially a columned style, and it must be remembered that some of the earliest vaults in existence are to be found in Egypt.[1] These last are, however, quite exceptional, and, as a general rule, the dwellers by the Nile used columns in such profusion as to justify our taking them as a characteristic Egyptian form, while some of the most splendid remains of this architecture consist of immense halls, like the famous Hypostyle Hall at Karnak, supported on range after range of mighty pillars, which bear up the stone slabs of the roof. Quite different are the ancient buildings of Mesopotamia. In that region neither timber nor building stone was plentiful, and the traditional material, mentioned as early as the account of the erection of the tower of Babel,[2] was brick, either sun-dried or less commonly burnt

---

[1] Lepsius, *Denkmäler*, Abth. 1. Taf. 89, and Dieulafoy, *L'Art Antique de la Perse*, Paris, 1884, etc., pt. iv. p. 21.   [2] Gen. xi. 3.

in the kiln. The researches of M. Place[1] have laid bare for us the ground-plan of a great Assyrian palace constructed in this traditional style, and his discoveries, together with the representations of buildings on the sculptured slabs in our museums, enable us to form a good idea of its elevation. The apartments of this palace of Sargon at Khorsabad were long and narrow, and were divided by walls of immense thickness, built of crude brick, and lined on the inside with a coating of white stucco. There is no trace of the employment of columns in the construction, nor any sign of the use of timber, and it is the belief of M. Place that the spaces were all covered by barrel vaults of sun-dried brick. This can hardly be said to be proved, but M. Place is able to produce in support of his view an actually existing vault of the kind, covering the passage of a gateway through the town walls hard by the palace. It is clear, moreover, from passages in ancient writers[2] that these brick vaults were actually used in dwelling-houses, as they are indeed to this day in the same regions, and the famous hanging-gardens of Babylon were borne up on vaults of a more substantial character.[3] This is sufficient to show that if we look to Egypt for the origin of the columned style, we must seek for some of the earliest examples of a consistent use of the vault in the antique land of Shinar.

With regard, next, to the employment of these forms in classical architecture, it will be seen at once that the Greek was essentially a columnar style, while the Romans had a special fondness for vaulting. Beside, however, the two styles popularly known as Greek and Roman or Tusco-Roman, classical architecture included a third, which may be termed *Hellenistic*, the style prevalent in the Asiatic and African cities founded

---

Place, *Ninive et l'Assyrie*, Paris, 1867.

[2] Strabo (xvi. 1, § 5) says of Babylon that all the houses were vaulted on account of the scarcity of timber. [3] *Ib. l.c.*

by Alexander and his successors, and closely connected, in all probability, with the architecture of Imperial Rome. Now the buildings of old Greece were almost without exception constructed by some modification of the favourite Hellenic form, the portico or colonnade. The Greek temple was from the first surrounded by colonnades, and these, with their superstructure, really formed the building, for the *cella*, or enclosed part within, had so little architectural character, that there was no natural way in which it could be lighted. Colonnades formed the chief secular buildings of a public kind that were to be found in the cities, and the shelter they afforded from the sun and the rain was all the interior accommodation required. Colonnades flanked or surrounded the market-places, and formed the localities of meeting for business in the neighbourhood of quais. The gymnasia, and similar places of public resort, consisted only in an enclosed space of ground, approached through a handsome portal, and provided within with colonnades and open *exedræ*. The private houses were built round courts, the porticoes of which were the real living apartments, the rooms proper being small and inconvenient. The theatres were not buildings in the strict sense at all, as they were formed mainly by the natural features of the ground, and were open above to the sky. Handsome colonnades were, however, considered a necessary adjunct.

In the Hellenistic period, after the conquests of Alexander, Greek architecture seems to have developed new forms. The portico still remained, it is true, its characteristic feature. Colonnades on a magnificent scale lined the streets of cities like Alexandria and Antioch, and the splendid Hellenistic structures connected with Herod's temple consisted almost entirely of porticoes. There is evidence, however, that the builders of these new cities made use of other architectural forms than the old. Greek column and architrave, and employed especially the arch and the vault, the use of which we have seen to be traditional in certain parts of the East. The architects of the Attalid sovereigns of Pergamon, in the third and the second century B.C., were particularly accomplished in stone vaulting, and on their work Professor Adler remarks: 'the construction of barrel vaults, and the transition from them to cross vaults built of regularly cut stones, had reached a high state of perfection in these Asiatic districts as early as the third, or certainly as the second, century before Christ.'[1] Of Alexandria it was said that the city was safe from fires, because no timber was employed in the construction of the houses, which were, on the contrary, roofed with vaults of rubble masonry.[2] The almost complete disappearance of all remains of the architecture of these magnificent Hellenistic cities is one of the most extraordinary facts in the history of building, but it would be a great mistake to ignore that architecture as a special and important factor in the development of construction. For there can be little question that it was from these cities that the architecture of imperial Rome was in great part derived. The conditions under which building was carried on within them were in many respects wholly unlike those which surrounded the architect of old Greece, and resemble much more nearly those existing in imperial Rome. For example, some Hellenistic cities, notably the great Seleukeia on the Tigris, arose in regions where it was difficult or impossible to obtain the building stone necessary for columns and architrave blocks,[3]

[1] Curtius, etc., *Beiträge zur Geschichte und Topographie Kleinasiens*, Berlin, 1872, p. 56.   [2] Hirtius, *de bello Alexandrino*, c. 1.

[3] Greeks of an earlier age had been placed in similar circumstances. As early as the seventh century B.C. they established themselves at Naukratis, in the Egyptian delta, and the exploration of this site, which is now proceeding under the able direction of Mr. Flinders Petrie, is of great value as showing how Greeks would build when deprived of their accustomed materials. Both timber and stone would only have been procurable with difficulty, and the town is accordingly built of the mud

so that brick and plaster, or rubble masonry, the characteristic Roman materials, had to be largely employed. The monarchs whose dynasties succeeded each other so rapidly on the Hellenistic thrones, needed architects who could build quickly and build for show, coating hastily-reared structures of base material with veneers of fine marble. Vast and splendidly-decorated interiors, suited for the use of a luxurious population, were also, we may well believe, a necessity of the time, and in these ways a preparation was made for the great development of an architecture of luxury and display in imperial Rome.

Roman architecture, as we see it for example in the Pantheon, was an architecture of interiors, and depended greatly for the effect of these on the use of vaulting. Perhaps the most characteristic of the buildings which made the glory of Cæsarian Rome were the gigantic Thermæ or bathing establishments, a glance at which will show the difference between the open-air architecture of old Greece and that which had been developed on Hellenistic and Eastern models in the Rome of the Emperors. These Thermæ were simply an extension on a vast scale of the old Greek gymnasium; but whereas that consisted only in an open space of ground with a few colonnades, the Roman Thermæ embraced, beside a beautifully laid out park, a grand complexus of splendid halls suited for bathing or for recreation. Though external grandeur was also aimed at, these buildings were essentially interiors, and derived their special architectural character from the mode of their construction. It is in them that we find the Roman love of vaulting exercised to its fullest extent. The apartments to

bricks of the neighbourhood. The sacred enclosures of the temples are not surrounded by porticoes, but by mud-brick walls. The houses seem all to have been of the same material, but as yet there is no evidence to show how they were roofed. Further architectural results from this most important 'find' will be awaited with much interest.

be roofed were of two forms, circular and oblong, and their dimensions were such as to require domes and vaults vaster in extent than any which have been attempted before or since. In the Pantheon, and in the transept of S. Maria degli Angeli at Rome, we have still remaining the finest examples of Roman vault-construction. Scientifically put together, with ribs of brick and filling in of concrete, the dome of the Pantheon embraces a space of more than 143 feet diameter, while the transept of S. Maria, originally a hall of the Thermæ of Diocletian, is covered with three cross vaults of 82 feet in span, which exceeds the span of the widest Northern Gothic nave, that of Chartres, in the proportion of 8 to 5.[1]

The architect of the fourth century had accordingly the choice of two styles of building. He could on the one hand confine himself as the old Greeks had done to combinations of the colonnade, producing buildings of simple construction with vertical supports and roofs of timber, or he could exercise himself upon structural problems connected with the use of the arch and vault. The Christian architect, be it clearly understood, adopted both of these styles. He did not by any means, as has sometimes been said of him, confine himself to unambitious undertakings, but boldly grappled with the difficulties of vault construction, in regard to which his pagan predecessor had set him such a brilliant example. It was only, however, with one form of the Roman vault (as far as regards work on a large scale) that he concerned himself, that is to say, with the cupola. His round or polygonal buildings received from the fourth century onwards the architectural glory of the dome, and the development of these buildings is the theme of the following chapter. For

---

[1] There are wider Christian vaults than those on the normal Gothic churches of the north. Gerona in Spain has a vault of 73 ft. The Cathedral of Angers (Romanesque) measures 54 feet.

the roofing of oblong spaces he did not attempt the system of cross-vaulting employed on such a magnificent scale in the halls of the Thermæ. The Christian Church had to wait till the eleventh century before it received a vault of masonry. During the whole of the early Christian period the builders of the churches contented themselves with a simple columnar construction and with a timber roof, and in this they followed the example of the builders of the pagan basilicas.

We return now to the question asked above, What was the Basilica?

The Basilica was nothing more than an ingenious arrangement of colonnades put together so as to form a commodious and well-lighted interior. Starting from the simple form of the covered portico, the favourite Greek building, we may trace the development of the basilica in the following manner.

In the early days of ancient republics, all the traffic and barter of a city, all the intercourse of burghers discussing politics or public news, all the legal disputations before magistrates, as well as many of the popular shows and amusements, went on in the public place, the Forum or the Agora. In a busy city it would be convenient to set apart certain spaces for the transaction of particular kinds of business, and the Roman bankers congregated about the Marsyas statue in the Forum, just as in our own day at Liverpool the 'Flags' accommodate the brokers of cotton. For further distinction from the common ground of the city, such a space might be surrounded by covered porticoes under which shelter from the weather could be obtained, while their roof supported a terrace, affording an agreeable and airy promenade. If the porticoes were bounded on the outside by a wall, further privacy would be obtained, and the result would be something resembling in ground-plan the Royal Exchange of London. The roofing over of the central space (which has now been carried out in the last-mentioned building) would be the last step necessary to change the portion of open pavement into a regular and commodious building. The roofing of the central space imparts however at once a new architectural character to the edifice.

We are dealing, it will be remembered, with a columned structure, for which a vault of masonry would be architecturally unsuitable. A vaulted basilica is indeed a possible form, and the churches of the eleventh and following centuries present us with examples of it. In Roman times, moreover, we have as an instance of such a structure the so-called basilica of Maxentius or Constantine (formerly known as the Temple of Peace) at Rome. This however was, like the Romanesque Minster, not a columned structure at all, but one of more massive character, derived from the vaulted halls of the Roman Thermæ, rather than from the class of buildings now under consideration. The basilica proper could support upon its columns nothing more formidable than a roof of timber, and the question how to place that roof so as to provide for the due admission of light into the interior was the crucial problem in basilica construction. The basilica as a hall of business was a place for crowds to come and go, for action, and for the strife of tongues; it was to be as much like the open air as possible— in fact it was nothing more than a portion of the public Forum, with the addition of shelter and comparative freedom from interruption. Abundant light was above all things necessary.

The architect of the basilica solved the problem before him in masterly fashion, and created a type of interior construction which, as adopted by Christian builders, has ruled the ecclesiastical architecture of western Europe through all the ages of Christianity. He preserved the porticoes with their terraced promenades all round his central space, and contrived to secure support for its roof without interfering with them. This was done by erecting a row of columns at the edge of this terrace, all round the central space, with a breastwork between them. This row of columns supported an entablature, and upon this rested the beams of the roof (Fig. 15). With this arrangement abundance of light would pour in through the columns of the upper arcade, while the

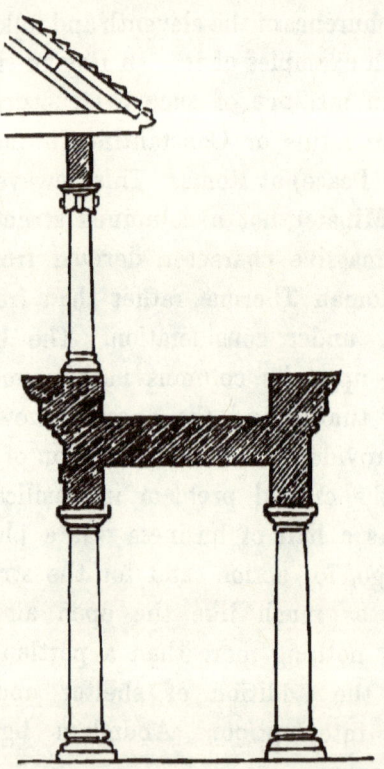

FIG. 15.—Section of part of Basilica in its (conjectured) earliest form.

roof would preserve the central space from the rain. A modification of this plan enabled the building to be more completely enclosed. The terrace could be covered in by carrying the exterior wall up above its level to a certain height, and constructing pent-house roofs leaning against the entablature of the upper columns, as in Fig. 16, and would then become a gallery running round the central space, and opening into it through the arcade. The light excluded in this way would be readmitted by raising a wall above the second entablature, and piercing it with a row of

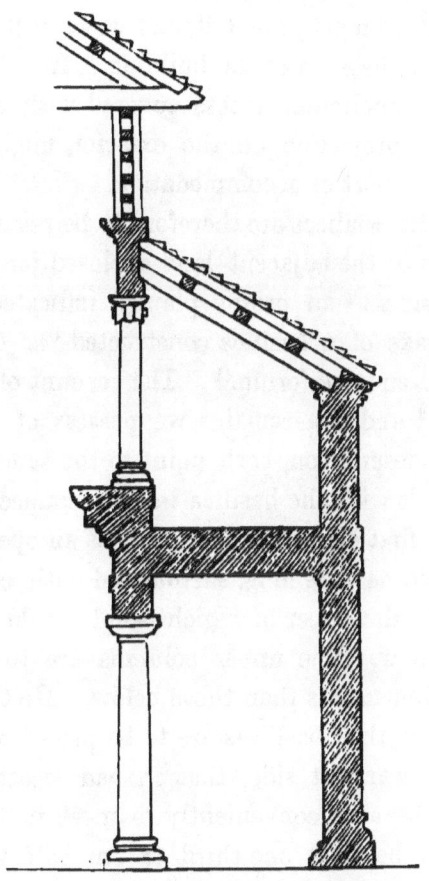

FIG. 10.—Section of part of Basilica. Normal form.

large windows, above which again would come the roof. The external walls of the galleries might also be pierced with windows. Entrances as numerous as necessary might be opened in the external wall of circuit, and adorned with columned porches, while the high-pitched roof would show a handsome gable at each end. Of the interior arrangements one only need be mentioned. In cases where part of the basilica was used as a law-court, a tribunal, or elevated stage for the seat of the magistrate, would be necessary. Sometimes, as in the case of the basilica which has been discovered at Pompeii, the tribunal was a rectangular platform at one of the ends of the building. In other instances an apse, or semicircular recess covered with a half dome, and forming a projection on the exterior, might be used to afford the same sort of accommodation.

The ancient basilicas are therefore to be regarded as parts of the forum or the adjacent land enclosed for various purposes of business—an origin plainly indicated by Cicero when he speaks of a basilica constructed '*ut forum laxaremus*'—' to extend the forum.'[1] The account of the basilica in Vitruvius,[2] and the remains we possess of actual buildings of this description, both point to the same conclusion. Vitruvius deals with the basilica in close connection with the forum. He first describes the forum as an open space two-thirds as wide as it is long, surrounded with colonnades of two stories, in the upper of which people could sit to survey the scene below. The upper columns are to be a fourth part less in dimensions than those below. He then proceeds at once to say that basilicas are to be placed adjoining the fora on the warmest side, that in bad weather the merchants may be able conveniently to meet in them. Their width should be from one-third to one-half their length; their columns as high as the width of their side colonnades, which again is to be one-third of that of the central space. The upper columns are to be less than the lower in the proportion given above, while between them is to run a breastwork to prevent those *walking on the roof of the porticoes*[3]

---

[1] *Ad Atticum*, iv. 16.   [2] *De Arch.*, v. 1.
[3] '*Supra basilicæ contignationem ambulantes*,' Vitruv. *l.c.*

(in the galleries?) from being seen by the merchants (gathered below).

The remains of actually existing basilicas are very scanty. Of the grandest of all, the Basilica Ulpia in the forum of Trajan, there are to be seen portions of the lateral colonnades which divided the area into a nave and four aisles. One of the ends, which are still unexplored, is represented on a fragment of the Capitoline plan of Rome (Fig. 17),[1] which shows not only a double but a triple row of columns passing across the narrow side of the building, outside of which, and probably therefore not connected with the basilica at all, appears a magnificent exedra or apse. The basilica at

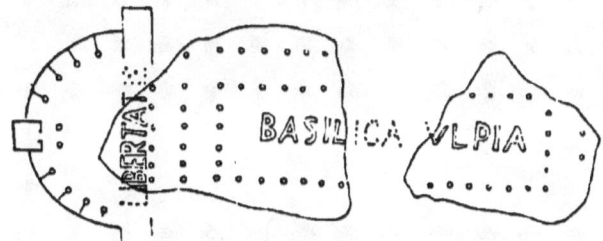

FIG. 17.—Basilica Ulpia from Capitoline plan.

Pompeii, described and illustrated in Professor Overbeck's work on the buried city,[2] possessed a single colonnade surrounding on all sides the central area, and was enclosed outside by a wall. Of most of the great basilicas of Rome no traces are now to be found, but the basilica Julia in the Roman Forum has been fortunately laid bare, and the plan of it affords the best illustration possible of the account of the basilica given in the text. This edifice, begun by Julius Cæsar, and completed by Augustus, lay along one side of the Forum, and covered an area of about 350 by 160 feet. As will be seen on the accompanying ground-plan, Fig. 18, it had a long and narrow central space, surrounded by double

[1] Jordan, *Forma Urbis*, Taf. iii.   [2] Overbeck, *Pompeji, etc.* 4 Aufl. p. 142.

porticoes forming its side aisles. There was no outer wall round the building, and the façade, of which we can form a notion from an ancient bas-relief, resembled the elevation of a Roman amphitheatre like that at Nîmes or Verona, or like the Colosseum, and consisted of two stories of round arched arcades supported on massive pillars. Between these pillars there was free passage in and out for those who had business in the basilica, or for mere loungers and spectators. We know that courts of law were held in the building, and

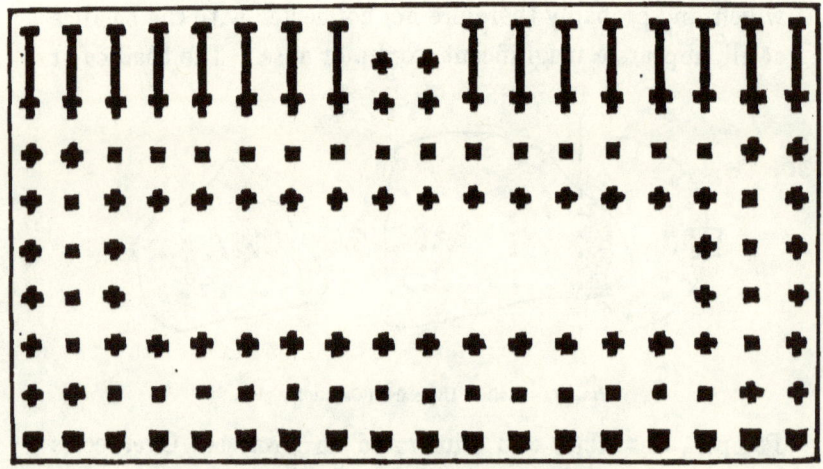

FIG. 18.—Basilica Julia, from Dutert, *Le Forum Romain*.

that the galleries which opened into the central space were frequented by those who wished to see—they were too far off to hear—the proceedings below.[1] Above the gallery there was a promenade on the roof, from which, according to Suetonius, the Emperor Caligula amused himself by throwing down gold coins to the crowd in the Forum beneath.[2] At the back of the structure, on the side away from the Forum, was apparently a row of shops, some of the walls of which are still to be seen *in situ*, and these, by opening directly

---

[1] Plin. *Ep.* vi. 33.   [2] *Calig.* 37.

into the basilica, made it still more like the open Forum. The law courts, with the busy crowd of suitors and minions of the law, would have reminded us of old Westminster Hall, while the shops and the loungers recalled the Palais Royal at Paris.

A further question, to which it is only possible to refer in passing, concerns the architectural history of this basilican construction; the source, that is, whence was derived the system of clerestory lighting, wherein its peculiarity consists. A review of the various columned halls of the ancient world cannot be attempted here, but there is one form of these which presents itself with a special claim upon the attention. This is the columned or 'hypostyle' hall, already mentioned as forming an important part of the Egyptian temple. To bring this into connection with the basilica, we have only to take note of an already quoted passage from Vitruvius, wherein he applies the name 'Egyptian' to a certain kind of private hall resembling a basilica. Now why should a columned hall, lighted by a clerestory, be called 'Egyptian'? The answer is that some of these temple-halls of Egypt contain the earliest, and in some ways the most effective, examples of this very system of interior lighting. The grandest structure of all, the hypostyle hall at Karnak, is supported upon 134 colossal columns, so arranged that the two centre rows, composed of far larger columns placed at greater distances apart, rise high above the roof of the side portions of the building, and bear a flat covering of their own. Between this covering and the main roof there is accordingly a considerable space, and this, filled in with marble lattice-work, is employed to give light to the interior.[1] Here we have the principle of the clerestory fully understood, and used to

[1] Lepsius, *Denkmäler*, Abth. i. Bl. 80. Perrot et Chipiez, *Histoire de l'Art dans l'Antiquité*, Paris, 1882, etc. *L'Egypte*, Pl. v.

produce a very striking and beautiful interior effect. 'No language,' writes Dr. Fergusson of this hall at Karnak, 'can convey an idea of its beauty. . . . The mass of its central piers, illumined by a flood of light from the clerestory, and the smaller pillars of the wings gradually fading into obscurity, are so arranged and lighted as to convey an idea of infinite space. At the same time the beauty and massiveness of the forms, and the brilliancy of their coloured decorations, all combine to stamp this as the greatest of man's architectural works.'[1]

But it may be asked, What possible link of connection can there be between ancient Egyptian temple-halls and the basilicas and festal apartments of the Romans of the Empire? A moment's consideration of the position and artistic importance of Alexandria will show that there was no source from which the Romans could more readily have borrowed an architectural form than Egypt. Alexandria, the metropolis of Hellenistic architecture and art, taught the Romans not a few lessons in building and decoration, and the '*Ægyptius œcus*' of the Roman house, as described by Vitruvius, was in all probability an Alexandrine structure borrowed by the ingenious and receptive Greeks of the Ptolemaic kingdom from the ruined halls of the valley of the Nile.

We should probably, moreover, not be mistaken in going a step further back, and conjecturing a derivation of the basilica form in general from the same ancient source. The communication of Greece with Egypt did not date from the founding of Alexandria, but had already existed for centuries, and the recent 'find' at Naukratis shows it to have been more close and constant than had previously been supposed. It is more than possible that the Egyptian clerestory gave to the builders of the columned halls of Greece the first idea of

---

[1] *History of Architecture*, i. 119.

that system of construction which was perfected in the magnificent Roman basilicas of Cæsar and of Trajan. If this conjecture is well founded, we may look with a new interest upon the hoary monuments of the Nileland, and may find amongst them the first beginning of an architectural system consecrated to us through its employment by forty generations of Christian builders.

. A general idea has now been obtained of the ancient basilica as it actually existed, and not as it has been reconstructed upon an assumed exact resemblance to the church of the Christians. The theory regarding that exact resemblance rests on a *petitio principii* which has been often exposed. The form of the Christian basilica was well known from existing examples. It was assumed that this was similar to that of the pagan buildings of the same name, and the latter were accordingly reconstructed upon the model of the former. This having been accomplished, writers could expatiate freely upon the resemblance of the two, and hold it proved that the Christian form of the structure was derived from the pagan. When the matter is looked at critically, however, it is seen that there were essential differences between the buildings which rendered exact copying altogether out of the question. In other words, the pagan basilica had to be modified in most important respects before its excellent architectural qualities could be made use of by Christian builders, or, to put it more accurately, the latter only required to borrow from the pagan basilica certain special features. What was the true relation of the two buildings under consideration will be seen when we proceed to examine the actual form of the Christian church. Before, however, we pass on to this subject, the reader's attention must be for a moment directed to an important fact which now presents itself. *This basilican type of interior was actually adapted to a form of congregational worship closely resembling that of the Christians at a period long before there can have been any question of a Christian basilica.* In other words, the *Jewish Synagogue* preceded the Christian Church in employing, at any rate occasionally, the basilican style, and from this fact a valuable light is thrown upon the architecture of the period on which we are engaged.

The history and constitution of the ancient Jewish Synagogue have been fully elucidated by Dr. Ginsburg in his article on the subject in Kitto's *Cyclopædia:* all that we are concerned with here is its architectural structure. Upon this subject we have two sources of information, literary and monumental. The remains of certain ancient synagogues, dating back to the earliest centuries of our era, are still to be seen in Galilee, and they have been recently thoroughly explored and measured in connection with the Survey of Western Palestine.[1] There exists besides a vast body of literary material in the Talmud and other writings of the Jews, with the accounts of early travellers like Benjamin of Tudela, from which information can be obtained both as to the normal and orthodox form and arrangement of the synagogue, and the special characteristics of certain prominent specimens.

The ruined synagogues of Galilee seemed at first to promise most interesting results alike for the student of architecture and of the Bible, but subsequent consideration has modified the view taken of their importance.[2] It was

---

[1] Survey of Western Palestine, Special Papers, p. 295 ff., esp. the paper by Lieutenant (now Lieut.-Colonel) Kitchener.

[2] There are notices of these synagogues scattered about the pages of the publications of the Palestine Exploration Fund, *e.g.* Memoirs, vol. i.; Quarterly Statements, 1869, p. 37; 1875, p. 22; 1877, pp. 118, 179; 1878, pp. 25, 32, 123. These are again collected in the 'Special Papers' already referred to.

thought, on the one hand, that some among them might be the actual buildings hallowed by the presence of Christ Himself and His disciples, and on the other, that we might argue safely from them to the normal form of the synagogue generally. The paper upon them by Colonel Kitchener[1] places them in a different light. The number of undoubted examples, according to his report, is eleven, and the whole area in which they occur is not much larger than Rutlandshire. They are a local phenomenon, and are curiously alike in form and in architectural treatment. He is disposed to believe that they date between the years 150 and 300 A.D., and that they were erected under Roman influence, if not by the hands of Roman legionaries. On this hypothesis can be best explained some curious features presented by these structures which seem to run counter to Jewish law and Jewish tastes. In considering them accordingly we are not, in his opinion, to imagine that we have before us the normal form of the orthodox Jewish meeting-house.

It is not indeed inherently likely that any one normal form existed, though in certain features of arrangement all synagogues would be found to agree. The abundance of these synagogues, existing as they did often in considerable numbers wherever there was a community of Jews,[2] and the wide area over which they were scattered, together with the known absence from among the Jews of a distinct national style of architecture, render it probable that there was no

[1] Special Papers, l.c.
[2] *E.g.* there were thirteen synagogues at Tiberias alone (*Babylon Berakhoth*, 8 a, Schwab, i. 252). The different sections of the Jewish community at Jerusalem had their distinct synagogues (Acts vi. 9), and the Jerusalem synagogues numbered, according to a somewhat doubtful tradition (*Jerusalem Megilla*, iii. 1, Schwab, vi. 235), no fewer than 480. Among these there is mentioned in the Talmud (*der Traktat Megilla*, translated by Rawicz, Frankfurt, 1883, p. 93)—if we may trust the rendering—a *synagogue of the smiths*, which is interesting as a Jewish parallel to the pagan *collegia* with their *scholœ* or lodge-rooms.

fixed type of construction recognised. To begin with, it was not necessary for the synagogue—or, as it was sometimes called, *e.g.* by Josephus,[1] *proseucha*, 'place of prayer,'—to be in the strict sense a building at all, for the 'place of prayer' of Acts xvi. 13 was evidently only an appointed station where the people gathered for service in the open air. We read of such suburban *proseuchæ* on the seashore,[2] or, as here, by river banks, where there would be provision for the needful ablutions. Epiphanius tells us that there were in old times places of prayer outside the cities both among the Jews and Samaritans, and mentions one at Neapolis, the ancient Sichem, in the plain without the walls, arranged like a theatre, and open to the sky above.[3] The synagogue, again, might consist in an enclosed space without a roof, in an open court, that is, surrounded by colonnades after the pattern of an early Mohammedan mosque. An interesting synagogue of this kind at Aleppo is described by a traveller of the seventeenth century. 'Early in the morning,' writes della Valle,[4] 'I went off to see the synagogue of the Hebrews, which has in Aleppo a high reputation for beauty and for antiquity.' Behind some buildings which masked it from the street 'we found the synagogue, which consists in a square open court of considerable size, with covered porticoes all round supported on double columns arranged in good architectural taste. On the right hand as one entered, beside the covered portico there is also a large hall which serves for holding the service when the weather is cold or rainy, while the open court is used when it is fair above.' In the middle of the court was 'a small cupola supported on four pilasters, beneath which, on a lofty and handsome

---

[1] *Life*, § 54.
[2] Josephus, *Antiq.* xiv. 10, § 23.
[3] Epiphanius, *Adversus Hæreses*, iii. Hær. 80.
[4] *Viaggi*, Roma, 1662, iv. 424.

platform, resembling the altars of Christian churches, were preserved the rolls of the law, and whence also the service was conducted.'

In general, however, the synagogues would be regularly enclosed and roofed edifices, provided with seats for the use of the congregation, and with fittings and furniture for the performance of public worship, and for the other purposes which the buildings were required to serve. Those of smallest size might be mere chapels, without internal colonnades, of the kind shown in Fig. 6, p. 54. This is taken from an early Christian mosaic in S. Maria Maggiore [1] at Rome, where it does duty for the temple at the door of which Zacharias is receiving the angel's message. The lamp suspended from the roof—a constant feature of the synagogue—makes it not unlikely that the artist was copying here a building actually used for Jewish worship at Rome. Synagogues of more extended size would be furnished with interior columns for the support of the roof,[2] and of this columned type are the synagogues of Galilee already mentioned. These were simple stone buildings of moderate size, the largest of the group, that of Meiron, measuring in the interior about 90 by 44 feet,[3] and are thus described in the words of Sir Charles Wilson:[4]—'The buildings are always rectangular, having the longest dimension in a nearly north

---

[1] Ciampini, *Vet. Mon.*, i. p. 200.

[2] A story in the Talmud (*Jerusalem Berakhoth*, v. 5, Schwab, i. 108) tells how a certain Doctor placed himself behind a column of the synagogue to escape notice. Again, 'R. Amé and R. Assa say: Although there were thirteen synagogues at Tiberias, it was fittest to pray *in the midst of the columns* where study was going on.'—*Babylon Berakhoth*, 8 a, Schwab, i. 252.

[3] The interior dimensions of the synagogues of Galilee measured by the Palestine explorers are as follows (omitting fractions):—Meiron. 90 × 44; Tel Hum (Capernaum?), 74 × 56; Kerazeh, 74 × 49; Kefr Birim, large synagogue, 60 × 46, small, 48 × 35.—Special Papers, p. 300.

[4] Quarterly Statements, ii. p. 37, and Special Papers, *l.c.*

and south direction, and the interiors are divided into five aisles by four rows of columns, except in the small synagogue of Kefr Birim, where there have been only two rows of columns and three aisles. The masonry of the walls is well built and solid, of native limestone, the stones are set without mortar, the beds and joints being "chiselled in" from two to five inches, and the remainder rough picked. The exterior faces are finely dressed, but the backs are left rough, more readily to take the plaster with which the interiors seem to have been covered, and of which some traces remain at Tel Hum.' The floors are paved with slabs of white limestone. The roof appears to have been formed of flat slabs supported on the numerous columns which were placed very close together, and to have been covered above by a layer of earth. On the lintels of the doors, which were usually three in number and at the southern end, there was some carving of a simple kind, representing objects familiar in Hebrew tradition, such as the vine, the pot of manna, Aaron's rod, the paschal lamb and the 'lion of the tribe of Judah,' and in one instance the Roman eagle. The large synagogue at Kefr Birim had in front of it a covered porch, and was decorated on the outside by pilasters. The style of these buildings (of which very little now remains), their masonry, mode of roofing and ornamentation—seems to resemble that of the buildings of central Syria explored and figured by de Vogüé.[1] The chief points of interest about them for our purpose are, on the one hand, their divergence from orthodox Jewish tradition (in the matter of orientation, and in the animal forms, including even the 'Abomination of Desolation' itself carved upon them), which has led Colonel Kitchener to set them apart as exceptional; and on the other, the interior arrangement of the columns. These do not divide the plan into a nave and side aisles,

[1] *Syrie Centrale*, etc.

SYNAGOGUES OF GRANDER TYPE.

but into five or three aisles of equal width, an arrangement familiar in the later Mohammedan mosques, such as that at Cordova. The many-columned mosque is, however, supposed to be developed out of the simple columned court in which the earliest mosques consist, by the gradual encroachment of the covered porticoes on the open space. Here there is no sign of any such development, and these many-columned synagogues appear rather as copies on a small scale of buildings like the hypostyle halls of the Egyptian temples, or the grand Persian 'hall of the hundred columns' at Persepolis. As regards the lighting of these synagogues, nothing can be made out beyond the fact that there were small windows in the front of the large synagogue at Kefr Birim. On the whole, they do not present to us the idea of really commodious buildings for congregational assembly, or answer to the grand conceptions which we form of the larger Jewish meeting-places from sundry passages in Josephus and the Talmud.

Josephus, for example, speaks of a *Proseucha* (synagogue) at Tiberias as a vast edifice capable of receiving a large concourse of people,[1] but gives no details from which we may infer its form. That the form of these great synagogues was, in some cases at any rate, *basilican*, is proved by an important passage in the Talmud, describing the splendid Jewish house of assembly at Alexandria, which has been rightly called 'a noble example of Hellenistic architecture.'[2] The community of Jews at Alexandria was the largest and most important of all those in the cities of the dispersion, and must have included a large proportion of the million Hebrews who inhabited Egypt in the early days of the Roman Empire.[3] The wealth and comparative independence of this community, the artistic skill at its disposal, and its opportunities

---

[1] Life, § 54.
[2] Hamburger, *Real-Encyklopädie für Bibel und Talmud*, art. "Alexandrien."
[3] Philo, *adv. Flaccum*, ii.

of obtaining from abroad any needful materials for building and decoration, make it certain that this great Alexandrian synagogue, the erection of which has been assigned to the second century B.C., would be a structure of the first importance.[1] Accordingly, the much-travelled Jehudah ben Ilai, who flourished during the latter half of the second century A.D., was wont to say that 'he who has not seen the double colonnade ($\delta\iota\pi\lambda\hat{\eta}\ \sigma\tau o\acute{a}$) of the synagogue at Alexandria has seen nothing of the splendour of Israel.'[2] The Talmudic writer goes on to say, 'it was a very lofty basilica,[3] composed of colonnades one within the other, and contained sometimes an assembly of worshippers twofold the number of the Children of Israel who came out of Egypt. There were there seventy thrones of gold adorned with pearls and precious stones, for the use of the seventy elders, members of the great sanhedrim, and each one was placed upon a pedestal the worth of which was twenty-five myriads of golden denars. In the midst of the building was a platform of wood, on which stood the conductor of the synagogue worship. When any one rose up to take part in the reading of the law, the presiding official waved a banner to give notice to those placed at a distance, in order that all might be able to answer AMEN at the same moment; and for every benediction recited by the leader of service, the president again waved the banner. Notwithstanding such a crowd of persons, all were seated in exact order. There were distinct groups formed by the separate corporations of the trades, so

---

[1] The Jewish community in Alexandria was presided over by a functionary called Alabarch (Jos. *Antiq.* xviii. 6. 3). One Alexander Lysimachus was Alabarch in the time of our Lord, and presented a costly offering of gold and silver plates for the adornment of the doors of Herod's temple at Jerusalem.—*Wars*, v. 5. 3.

[2] *Jerusalem Soucca*, v. 1, Schwab, vi. 42.

[3] The technical term '*basilica*' is introduced into the Hebrew text of the Talmud. Schwab, vi. p. iv.

NORMAL PLAN OF THE GREATER SYNAGOGUES. 103

that when strangers (ξενοί) arrived they were able to enter into communication with comrades of the same trade, and in that way meet with work from which to support themselves. The edifice has been destroyed by Trajan the impious.'

Here is a picture which historians of architecture and students of Christian antiquities may both alike contemplate with advantage. We must allow, of course, for Oriental exaggeration in numbers and valuations, but there can be no question that the building destroyed by Trajan was a basilica of magnificent proportions and of extreme richness of decoration, in the fine style of Hellenistic architecture and ornament. Some words in the *Itinerary* of Benjamin of Tudela,[1] descriptive of the synagogue at Bagdad in the middle of the twelfth century, may be quoted as a supplement to what has just been given. 'The metropolitan synagogue of the Prince of the Captivity is ornamented with pillars of richly coloured marble, plated with gold and silver. On the pillars are inscribed verses of the Psalms in letters of gold. The ascent to the holy Ark is composed of ten marble steps, on the uppermost of which are the stalls set apart for the Prince of the Captivity, and the other princes of the house of David.'

Let us now, in the light of these descriptions, and of other passages in the Talmud and elsewhere, attempt to draw out the normal plan of a synagogue of the grander type, like that at Alexandria. Before the entrance to the building we may assume that there was a porch, as is the case in the existing example at Kefr Birim in Galilee, and seems to be demanded by the mention of devout doctors sitting reading the law at the gate of a synagogue.[2] Here would be provided water for the necessary ablutions. According to orthodox Rabbinical prescription, the building

---

[1] *Itinerary*, ed. Asher, Lond. 1840, p. 104.
[2] *Jerusalem Berakhoth*, v. 1, Schwab, i. 9S.

should be on the most elevated ground available, and should be oriented in the direction of Jerusalem. The following fine passage from the Talmud[1] conveys this prescription in notable language: 'Those who are in foreign countries, beyond the borders of Palestine, ought in praying to turn their face towards the sacred land, as it is written, "They shall address their prayer to Thee by the way of the land which Thou hast given to their ancestors" (1 Kings viii. 48). Those who dwell in Palestine direct their countenance towards Jerusalem, for it is written, "They shall pray unto Thee toward the city which thou hast chosen" (*Ibid.* 44). Those who make their prayer at Jerusalem turn towards the mount of the Temple, as it is said in the same verse: "And the house which I have builded in Thy name." Those who are upon the mount of the Temple turn towards the Holy of Holies, as it is said: "They shall address their prayer to Thee in this place, and Thou wilt hear it in heaven Thy dwelling-place, Thou wilt hear it, and wilt pardon" (1 Kings viii. 30). Hence it follows that those of the north should turn towards the south, those of the south towards the north, the men of the east towards the west, the men of the west towards the east, so that all Israel shall turn in the act of prayer towards the same place, as it is written, "My house shall be called a house of prayer for all the nations" (Isaiah lvi. 7).'[2] According to this prescription the head of the building would point towards Jerusalem, while the doors were upon the opposite side, and there, at the end of the interior space,

---

[1] *Jerusalem Berakhoth*, iv. 6, Schwab, i. 90.

[2] One of the reasons for considering the synagogues of Galilee exceptional is their orientation. They are all turned towards the north, and worshippers entering by the doors on the south would turn their backs, and not their faces, towards Jerusalem. It must be added, however, that this question of turning in prayer towards a single spot was somewhat controverted among the Doctors. There were not wanting those who said, in the spirit of Christ's words to the woman of Samaria, 'God is

visible at once to all who entered through the portal, was placed the receptacle for the sacred treasure of the synagogue, the ark containing the rolls of the Law. We are supposing that the basilican form would be that usually adopted for important synagogues. The mention of the 'double colonnade' at Alexandria shows that it was a basilica with nave and four side aisles, and this form seems the best to answer to the characteristics of the synagogue, as we trace them in the Talmudic stories. For it would give, first, abundance of light to the interior. That this was considered necessary follows from the great importance attached to the reading of the Law without any mistake on the part of the ministrant, and from the fact that the building was not only used for service but for the study of the Law.[1] It is also proved by the following passages. On the eve of the Passover it was prescribed that search should be made by candle light in all places to see if any leavened bread were lying about. 'R. Jeremia demanded if it were necessary to carry on this search in the synagogues and the halls of study. . . . The question was justified because, *by reason of the abundant light which pervades these interiors*, one might doubt whether it would not be sufficient to examine them by daylight without having recourse to a candle.'[2] Again, it was said, 'One ought only to pray in a house which has windows, as it is said (Daniel vi. 11), He had opened the windows of his chamber towards Jerusalem.'[3] It would seem natural from this to suppose that, in a basilica like this at Alexandria,

everywhere,' and deprecated this formal orientation. Again, it was doubted by some whether, after the destruction of the second Temple, the duty of thus turning was not at an end. Some Jewish teachers objected ever to turn towards the east, as in so doing they would be following the example of the heathen Persians.—*Jer. Berakhoth*, iv. 6, Schwab, i. 90; Hamburger, *Real-Encyk.*, art. 'Synagoge.'

[1] *Babylon Berakhoth*, 8 a, Schwab, i. 252.
[2] *Jerusalem Pesahim*, i., Schwab, v. 2.
[3] *Babylon Berakhoth*, 34 b, Schwab, i. 367.

the ark of the Law would be placed in an exedra at the Jerusalem end of the building, and that in the walls of this windows would be pierced, opening, like those of Daniel, towards the Holy City. Such a position for the ark would give it command of the whole interior, and it would accord also with another important feature of the synagogue. This feature was the range of 'chief-seats' for the rulers of the synagogue, and the doctors of the law. These, it is well known, faced the congregation, which accounts for the love of them on the part of the scribes and Pharisees.[1] Where could they be placed more suitably than in a semicircle on each side of the ark? for we cannot suppose that any doctor would actually turn his back upon the sacred receptacle. This is indeed the position which seems marked out for them by Benjamin of Tudela's words about the synagogue at Bagdad, which doubtless had preserved its form from Talmudic days. 'The ascent to the holy Ark is composed of ten marble steps, on the uppermost of which are the stalls set apart for the Prince of the Captivity, and the other princes of the house of David.'[2] If, therefore, we suppose a semicircular exedra, or apse, with steps ascending to it, we should obtain a form like that shown in Fig. 19, which is that of a basilica of the largest dimensions, with double aisles surmounted by galleries, but with the colonnade and gallery removed from the end opposite to the entrance, and their place taken by a commanding apse.[3] In the centre of this, at the extreme end of the edifice, would be placed, possibly in a recess in

---

[1] Matthew xxiii. 6.   [2] *Itinerary, l.c.*

[3] It may seem a somewhat daring conjecture to place an apse at the end of an Hellenistic building, but whence did the Romans derive this characteristic feature of their later architecture? There are no Roman apses earlier than those in the interior of the Pantheon, but there is a Greek one in the terrace wall of a temple at Pergamon, dating from at least as far back as the second century B.C. See *Ausgrabungen zu Pergamon*, 1880-81, Berlin, 1882, p. 29. The names used by the Romans for the apse—*exedra, hemicyclium, absis*—are Greek.

THE 'CHIEF SEATS.' 107

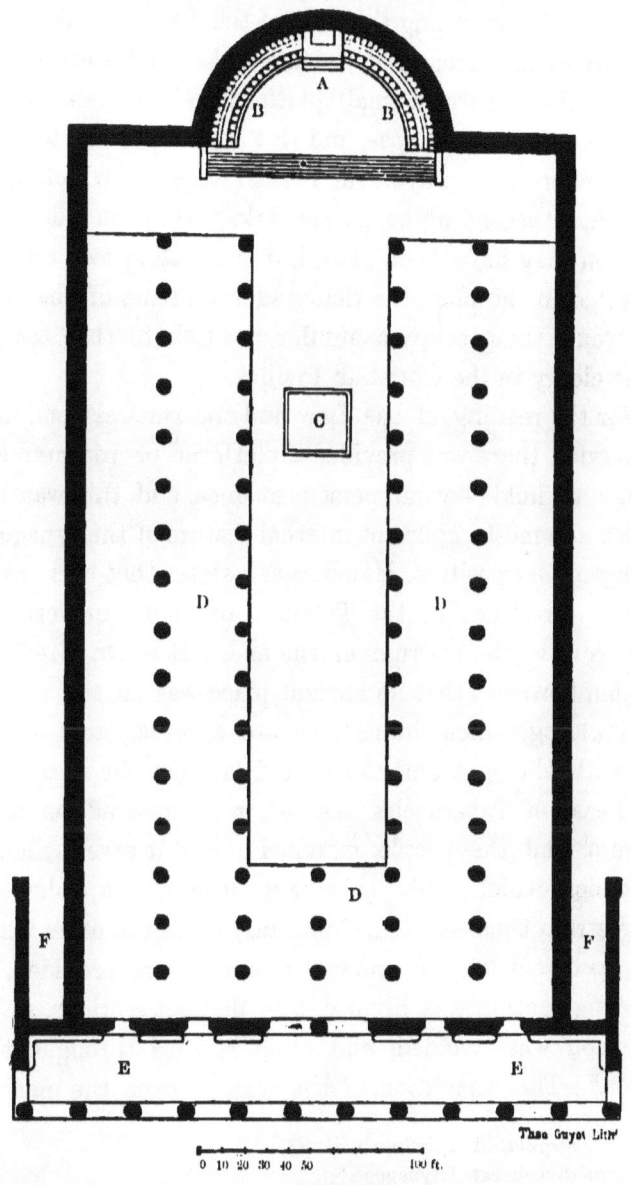

FIG. 19.—Conjectural plan of the Basilica at Alexandria, as an example of the Jewish Synagogue of the grandest type.

A. Ark containing the rolls of the Law.
B.B. 'Chief seats' for the elders.
C. Rostrum.
D.D.D. Gallery for the women.
E.E. Porch.
F.F. Stairs giving access to gallery.

the wall, and always with a rich curtain in front, the moveable ark of wood, containing the rolls of the Law, and sometimes, in another compartment, vestments and sacred utensils.[1] In front of the ark was a small platform, raised a step or two above the floor of the apse, and this was the appointed place for the offering of prayer, the regular term for which, in the Talmud, is 'ascending before the Ark.' On each side of this platform may have been placed, if needful in two ranks, as suggested in the plan, the richly adorned seats of the elders, who would thus occupy a similar position to that assigned to the clergy in the Christian basilica.

For the reading of the Law and the general conduct of the service, there was provided a platform or rostrum large enough to hold several persons at once, and this was, after the ark, the most important internal feature of the synagogue. As regards its position, Hamburger[2] states that there are no explicit directions in the Talmud, and most modern synagogues have the rostrum at the end. It is Dr. Ginsburg's opinion, however, that its ancient place was in the midst of the building, which, indeed, we are expressly told was the case with the rostrum at Alexandria. On the occasion of the Feast of Tabernacles the ark was removed on to the rostrum, and the people marched round it seven times in procession, which could only have been accomplished with the rostrum thus situated. With respect to the other fittings and arrangements, we know that seats were provided, and the congregation was divided into distinct sections, so that confusion was avoided, and order reigned throughout the edifice.[3] The separation of the women from the men is a

---

[1] *Jer. Megilla*, iii. 1, Schwab, vi. 237.
[2] *Real-Encyk.* art. 'Synagoge.'
[3] *Jerusalem Megilla*, iii., the *locus classicus* for the arrangements of the synagogue, contains a notice of its various fittings,—seats and benches, the curtain of the ark, the rostrum and the planks composing it, the reading desks, etc.

familiar feature of the synagogue, and for this provision could easily be made in a basilica by using the galleries characteristic of that form of building. Separate entrances might readily be contrived as suggested on the plan, and an arrangement would result similar to that which prevailed in Byzantine churches, such as S. Sophia. In the body of the building, in cases when, after the example at Alexandria, the members of the different trades sat in sections apart, the appointed location of each in the synagogue would answer the purpose of a kind of *schola*, or place of trade assembly. An interesting suggestion here offers itself. This custom at Alexandria has been brought into connection with the enforced temporary residence in Egypt of Joseph the carpenter of Nazareth.[1] It seems natural to suppose that he would wend his way towards the metropolis of the Egyptian Jews, the city which claimed for itself a monopoly of the trade and commerce of the country.[2] The special arrangements made at Alexandria for the reception of strangers, ξενοί (p. 103), would be an additional attraction, and we may picture him entering the immense synagogue on the Sabbath, and forgathering with the brethren of his craft, while Mary, with her infant in her arms, ascended among the daughters of her people into the gallery above.

The Jewish Synagogue is a building so interesting in itself, and so important from its affinities to the Christian meeting-house, that no apology is needed for the space here devoted to it, nor can we turn from it without one glance back at its general aspects as a religious and social institution. For the synagogue, the expression of all that was best in the national life in the period we are considering, was bound up in the most intimate way with the being of the Jewish communities. Around the synagogue, rather than the temple, moved the activity of the national party of the Pharisees, the party

---

[1] Matthew ii. 14. [2] Strabo, xvii. 1, § 13.

of the mild but princely Hillel, of the wise and noble Jehudah ben Ilai, of the liberal Gamaliel I., and his more famous disciple Saul of Tarsus. The Talmud is the literary expression of this side of Judaism, and from the Talmud we may learn not only the outward features of the synagogue, but the place it held, or was designed to hold, in the religious life of the people at large.

The synagogue was far from possessing the mysterious sanctity of the temple,[1] and knew nothing of the rigorous restrictions relating to Levitical impurity, which excluded all Jews not Levitically clean, and *a fortiori* all Gentiles, from the inner courts on Mount Zion. Hence, as already noticed, the mixed multitude was admitted to its services, but provision was at the same time carefully made that its own sacredness should always be preserved to it. It was never to be used, or even approached, without a due feeling of religious reverence. Its very stones were sacred from the moment they were cut in the quarry, and if the building fell into disuse its material ought not to be sold for secular purposes. Even ruined synagogues were still holy, and should be left as they were till their condition inspired a pious wish to rebuild them.[2] A secular hall might be dedicated as a synagogue, just as it might be made into a Christian church, and would become sacred so soon as the rolls of the law had been deposited in the place prepared for them.[3] The buildings were used for public service three or four times a week, and for daily private prayer, so that, while possessing doors with bolts,[4] they would stand in the daytime always open. Apart from divine service their chief religious use was for reading, study,[5] and disputation, which

---

[1] *Babylon Berakhoth*, 62 b, Schwab, i. 499.
[2] *Jerusalem Megilla*, iii. 1, Schwab, vi. 235.   [3] *Ibid.* p. 236.
[4] *Jerus. Eroubim*, x. 10, Schwab, iv. 302.
[5] 'R. Abaji said, Formerly I learned my Talmud at home, and went to pray in the synagogue, but after I had read the verse of the Psalm,

went on in the building itself when there were no siderooms dedicated to these purposes.[1] 'The synagogues,' said R. Josué b. Levy, 'are made for the use of the wise,'[2] and became the recognised home of the teacher and the student, as well in their hours of relaxation as of study.[3] It was, on the other hand, an offence for the ordinary person to stroll aimlessly into the building, or use it for trifling purposes. It was not to be employed as a short cut, a place of promenade or of shelter from the sun and rain: to sleep therein or to eat and drink was forbidden,[4] though strangers might be sheltered and provided with refreshment without any desecration.[5] To carry on commercial transactions in the synagogue was as bad as to turn it into a charnel house.[6] The following is an illustrative story:—' R. R. Rawina and Adi asked a question of Rawi in the street. There came on a sudden shower, and they took refuge in the neighbouring synagogue.' This was held to be rightly done, because it was not merely shelter that was sought, but the opportunity of continuing undisturbed a conversation on a question of the Law.[7] The synagogues were employed also for judicial proceedings: the local courts were held in them, and there also their sentences were carried out by the infliction of stripes.[8] Finally, funeral ceremonies were conducted in them, but only in the case of distinguished persons as the expression of general mourning.[9]

Lord, I love the habitation of Thine house, I learned my Talmud also in the synagogue.'—*Megilla*, trans. by Rawicz, p. 105.

[1] Every synagogue was also a hall of study, but separate apartments for study were (generally?) provided. The numerous synagogues at Jerusalem are said each to have had a hall for reading the Scriptures, and a hall for studying the Mischna.—*Jer. Megilla*, iii. 1, Schwab, vi. 235.

[2] *Ibid. l.c.* p. 240. [3] *Meg.* Rawicz, p. 102.
[4] *Ibid.* p. 103; *Jer. Meg. l.c.*
[5] *Jerusalem Pesahim*, i., Schwab, v. 2. [6] *Meg.* Rawicz, p. 102.
[7] *Ibid.* p. 103. [8] Matt. x. 17. [9] *Meg.* Rawicz, p. 102.

The synagogue appears, accordingly, as an essentially popular institution, democratic in the sense that no priestly caste or privileged family had part or inheritance therein, yet preserved in inviolable sacredness through the religious habits of the community. The chief influence in the preservation of those habits was education. 'The whole universe,' said the Doctors, 'has less value than a single one of the prescriptions of the Law;'[1] but they also laid it down that 'the act of hearing a passage of the Bible repeated by a little grandson was equal to the audition of the law on Mount Sinai.' In the spirit of this fine maxim the Jewish boy was brought up from the first to feel that in the life of the synagogue he should find his highest and holiest pleasure, and that it would some day be his part to lead therein the solemn service. So soon as he came to man's estate he might be called upon to take public control of the worship, to read without hesitation or blunder the sentences of the Law, to direct from the rostrum the singing of the congregation, or to give the sacerdotal blessing from before the ark.[2] We who are satisfied to set apart a certain number of our community and a certain measure of our time for the service of the sanctuary, hardly can realise what a powerful element in the religious education of an earnest Hebrew layman was the consciousness that he must qualify himself not only to follow but to lead the daily devotions of his people. That such was the office of the members of the community at large was the most open fulfilment of the promise which it was the highest aspiration of Judaism to realise, 'Ye shall be unto me a kingdom of priests, and an holy nation.'[3] Announced at the most solemn crisis of the national history,

[1] *Jerusalem Pea*, i. 1, Schwab, ii. 14.
[2] In *Jer. Soucca*, iii. 12, Schwab, vi. 31, it is stated that the youth is not to be permitted to do this before his beard is grown.
[3] Exodus xix. 6.

the opening of the drama of Mount Sinai, this promise had been kept alive through the influence of the prophets, who stood outside the temple system with its hereditary priesthood. The writer of the second of Maccabees[1] re-asserts the same principle, that God 'the Saviour of all his people, hath given to all the inheritance and the kingdom and the priesthood and the consecration,' and St. Peter appeals to it again in his epistle to the 'sojourners of the dispersion.'[2] There were two ways, however, in which the promise might be fulfilled. The prophets, in immortal words, had claimed for it a spiritual fulfilment by the righteousness of the heart, and Christ, in whom the prophetic line received its close and consecration, made this the keynote of His teaching. The Pharisees interpreted it in a legal fashion. They had emancipated themselves from the hereditary priesthood, and sought to elevate each individual man, through the exactest regulation of conduct, to a priestly order of a nobler type. Their holiness tended, however, as we know too well, to become an external matter, resulting in that formalism and pride which drew down the sharpest rebukes from the lips of our Lord. This fundamental error of Pharisaic teaching we may recognise to the full, without being blinded thereby to the elements of true moral strength in the orthodox Judaism of the hall of study and the synagogue. He is indeed an unsympathetic reader who can turn over the pages of the Talmud without discerning that underneath this Rabbinical casuistry and apparent zeal only for the 'letter which killeth,' there lay the deep foundation upon which the lasting greatness of the people is based, the foundation of personal self-respect and purity, of blameless family life, of infinite care for the nurture of the young, and above all an intense religiousness. 'An holy nation,' in the sense in which the words would be used by Christ, the

---

[1] 2 Macc. ii. 17.   [2] 1 Peter ii. 5, 9.

Jews were not, but their effort after such holiness as they discerned was still a noble one, and it was the greatness not the faults of the people that these synagogal services called into play.

We may picture to ourselves the broad spaces of the Alexandrine basilica flanked with inscribed and gilded columns,[1] and canopied with a roof curiously graven,[2] and rich with colour and glance of metal. On the rostrum in the centre, midmost of the congregation, stands the director of the worship, and summons now one and now another to take part in reading or in prayer. He who is asked at first refuses, as not deeming himself worthy of the proffered honour,[3] but, duly urged, comes forward to the desk, and receives from an attendant's hands the sacred roll of the Law. Practised as is his voice, it cannot penetrate to every part of the immense interior, and the presiding officer waves an embroidered banner to give the signal for the responses. There is movement about the doors as men come and go, while, calmly enthroned beside the ark, inflexible as the Law itself, in face of the thronging thousands, sit the elders of Israel.

Or let us imagine some great anniversary, when the ordinary service gives place to a solemn commemoration. How deeply stirred were those vast audiences with the joy and the sorrow and the dread, the recollection of the past, the hopeless longing, the far-off expectation, which the grandly-contrasted Jewish festivals called up to every mind. With what mighty waves of passion were they swept on the days of triumph, when the silver trumpets of remembrance pealed forth, and men bethought them of the blasts on Sinai, of the fall of the ramparts of Jericho, of the glories of the

---

[1] As at Bagdad, *ante*, p. 103.
[2] Like the roofs of Herod's porticoes round the temple courts.—Jos. *Wars*, v. 5, 2.
[3] *Bab. Berakhoth*, 34 a, Schwab, i. 364.

temple service, or, of nearer date, the heroic clarions which ring the note of faith and of victory through the first of the Maccabees. How dread the gloom, when, on days of fast and penitence, the people cast cinders on their heads and the sacred ark was moved from its place in the synagogue, and covered not as of yore with gold, but with the ashes of wood, was borne slowly out into the street, while the bones of the pious trembled with reverence and holy sorrow.[1]

Such were the scenes on which Mary of Nazareth may have looked down during those years of Egyptian sojourn. This was indeed to the outward eye the 'splendour of Israel,' and not without awe would she gaze on the assembly of the wealthy and powerful and learned of her race. What a contrast were they to the stranger fugitives from their obscure village of Northern Palestine! Yet perchance as she pondered once more over the sayings treasured in her heart, there may have flashed before her eyes a vision of a greater glory that should come, as, kindling again with the enthusiasm of the Magnificat, she knew that it was upon her the blessing of the generations would be spoken, while for the fall and rising again of these many in Israel should be set the unconscious babe which slumbered on her breast.

Turning now from the meeting-house of the Jews to that of the Christians, let us approach in thought a Christian basilica of the fourth or fifth century, and consider the form, arrangement, and decoration of its various parts, tracing wherever it is possible their origin in the earlier buildings of the ages of persecution.

The Christian church was occasionally surrounded by a

---

[1] *Jerusalem Taanith*, ii. 1, Schwab, vi. 152. 'R. Zeira said: Every time that I have seen the ark thus borne my body was seized with trembling.'

wall of enceinte, *peribolus*, marking out a sacred precinct, within which it stood with its subsidiary structures grouped about it. So stood the basilica erected by order of Constantine near the supposed site of the Holy Sepulchre at Jerusalem, and in such a precinct rose the grand church at Tyre, described by Eusebius in the tenth book of his Ecclesiastical History. This precinct, a distinct copy of the *temenos* or sacred enclosure of classical temples and of Mount Zion, cannot have been of frequent occurrence. Almost universal, on the other hand, was the important preliminary feature of the forecourt or *atrium* which lay before the entrance side of the building. A portico gives admission to this, and we find ourselves in a court enclosed on three sides by colonnades, and containing in the midst a fountain under a baldachin or canopy, called *Cantharus*. At this fountain those about to enter the sacred building washed their hands and lips in token of purification. The arcades round the court, and its central space, afforded room for groups to walk and to converse. It was as it were the Christian forum where the business of the community was transacted, and it has been considered by some that it was a copy of the public forum of the city, the church taking the place of the basilica adjacent thereto. It is far more probable that it was a reminiscence of the *peristyle*, or court surrounded by colonnades, of the private mansion, out of which opened the *œcus* and other festal apartments. In such open courts had been held the meetings of the early Christians for daily salutation, for council, for the reception of converts, and for informal business generally, and we cannot doubt that in the fountains which made pleasant music in their midst Christian baptism had been in the early days not seldom administered. In the *atrium* of the basilica we may see, therefore, a clear trace of the primitive meetings in private houses.

THE INTERIOR. 117

The fourth side of the court is occupied by the façade of the church, which at this period commonly lay with its entrance to the east, and with the altar at the western end. The arrangement of vestibules varied in different places. Sometimes the continuation of the colonnade along the fourth side in front of the church formed a porch before it, from which access was gained immediately into the building. Sometimes a special vestibule was interposed, or the portion of the church nearest the doors was marked off by barriers. These arrangements had reference to the different classes of catechumens and penitents of various degrees, who were not in full fellowship with the Church, but had their appointed places in the outer courts.

On entering the building through doors over which hung rich curtains, indicated in Figs. 6 and 12, we should have seen before us a light, spacious and well-proportioned hall of oblong plan, divided by rows of columns into a central space and two or four side aisles. Above the colonnades would possibly appear the galleries characteristic of the pagan basilicas, but more frequently—as in almost all the existing basilicas of Rome and Ravenna—a plain wall would rise above the columns of the nave, the upper portion of which would be pierced with windows many and large, admitting a flood of light into the edifice.[1] The roof might either be open, that is, with the timbers of the construction shown, as in the ancient example in S. Sabina on the Aventine at Rome, or concealed by a ceiling portioned out into squares—cassettes—and richly carved and gilded.

[1] The earliest basilicas had a profusion of windows, which in later times were diminished both in size and number. The notion of a 'dim religious light' is a refinement of comparatively modern times. The size of the clerestory windows of the basilica in Fig. 14 is remarkable. Abundance of light characterised, as we shall see, the churches of Byzantium.

Here at once is apparent the significance of the name 'basilica,' as applied to the Christian church. The question of church-building at this period must have been greatly a question of providing accommodation for the immensely increasing numbers of the Christians in the latter part of the third century, or in the days of Constantine, and the adoption of the basilican plan was the readiest means of procuring it. The importance of this plan lay in the fact that it allowed a far larger building to be erected and roofed at a moderate cost than was possible if the form of the simple hall or *schola* were maintained. These were necessarily limited in size through the difficulty of roofing a wide unbroken space. The contrivance of interior colonnades which might be two or four in number, enabled the building to be at once enlarged to the magnificent proportions of old St. Peter's at Rome, 380 feet in length, by a total width of 212, while the roofing, being divided into parallel sections over nave and aisles, was still perfectly easy. The lighting of the interior by windows in the upper walls of the nave was a natural consequence, and a Christian congregation might find itself in a building of basilican form before the name 'basilica' had yet been applied to it. The name itself came into use during the first half of the fourth century. There is some trace of its employment in documents relating to transactions in Africa during the Diocletian persecution of A.D. 303,[1] but before that time it was never employed, and this fact agrees with the view here put forward that the Christian church was enlarged into a basilica at the end of the third or beginning of the fourth century, and received the name in consequence of the architectural change. Previous to this the meeting-house of the faithful was called by such names as οἶκος προσευκτήριος, 'house of prayer,' οἶκος ἐκκλησίας, 'house of assembly,' *ecclesia*,

[1] De Rossi, *Rom. Sott.* iii. 461.

oratorium, or more commonly 'house of the Lord,' *domini-cum*, in Greek κυριακόν, whence, in all probability, the modern 'kirche,' 'kirk,' and 'church.' A certain pilgrim from Bordeaux, who visited Jerusalem in the year 333, in writing of the church built by Constantine near the site of the Holy Sepulchre, uses the words '*basilica*, ID EST DOMI-NICUM, *miræ pulchritudinis*,' which show that the term '*basilica*' was only then coming into general use, and needed to be explained by the more familiar '*dominicum*.' After this time *basilica* is the common word, and we may doubtless find a secondary reason for its prevalence in the connection of the term 'royal' with the kingly character of Christ. So far, then, as the arrangement of side aisles and clerestory lighting are concerned, these were undoubtedly derived, with the title of the whole structure, from the pagan basilica. Here, however, the resemblance ends, and we come now to architectural features in which the two 'basilicas' are forcibly contrasted.

The most prominent feature in the interior of the Christian basilica is the imposing apse in which the central nave terminates, and in which is the seat of the clergy and the altar. Few architectural effects are, indeed, simpler and yet more striking than the apsidal end of a great Christian basilica. In examples like S. Apollinare in Classe and the other churches at Ravenna, the columns stretch in unbroken line to the end wall which is hollowed out in the midst by the spacious apse, but in the grander Roman basilicas, like old St. Peter's and St. Paul's, the lateral colonnades end before the apse is reached, and high up across the nave is thrown a mighty arch, called the arch of triumph, one of the most conspicuous features of the edifice. Passing under this arch, we find ourselves in a free space before the apse extending the whole breadth of the church, or even, as in old St. Peter's, beyond the lateral walls, so as to form a regular transept. From this a flight of steps leads up into the apse, which is always raised above the level of the church, and is furnished with a stone seat round its circuit, broken at the central point by a raised chair or throne. Here was the post of the bishop, who had on each side of him the presbyters or elders, while the deacons and under-officers stood ready at hand for service, just as we have seen Paul the bishop sitting in the midst of his officers in the Christian meeting-place at Cirta.[1] The altar was placed in the front of the apse before the throne of the bishop, who stood behind it to officiate at the service, and would in this position face the east, towards which he would send forth his prayers. The celebration of the Eucharist in the cemeteries, under the circumstances at once solemn and exciting referred to in the preceding chapters, had created a connection between the altar and the tombs of the saints, which led to the construction in the churches of receptacles for relics beneath the holy table. These were the so-called ' *Confessiones*,' or small excavations to receive one or more sarcophagi, under the apse, which exist in several of the Roman basilicas, and have, as the origin of the crypt, such important significance for church architecture of the succeeding age.

The same reminiscence of the services at the cemeteries is held by some of the best authorities of the day to be the true explanation of the adoption in Christian architecture of the apse itself.[2] According to their view it is derived directly from the exedræ of the cemeteries, which have been noticed on a previous page.[3] It will be remembered how the priest stood beneath the semi-dome of these apses to

---

[1] *Ante*, p. 46.
[2] This is the view taken by Professor Kraus in his article ' Basilika ' in the *Real-Encyklopädie der christlichen Alterthümer*, p. 119.
[3] See *ante*, p. 56, and frontispiece.

perform the ceremonials for the martyrs, while the people looked in from without through the open side of the building. So much feeling was evoked by these ceremonials, that the popular taste might naturally crave for similar architectural surroundings when the assemblies were held in the regular churches of the ages of peace. If this view is correct, the apsidal end of the church of the fourth century becomes a memorial cella of the martyr whose bones repose in the crypt below, while the space in front represents the portion of the area of the cemetery where the congregation used to gather, the only difference being that it is now enclosed and covered in, just as the open spaces of the city were enclosed by the public basilicas. A glance back at Fig. 11, which shows a basilica actually, as it were, growing out of a memorial cella, will strikingly enforce this view.

For this theory there is undoubtedly much to be said. It emphasises the thoroughly monumental character of the apsidal end of the church, and exhibits it as designed to glorify, by architectural grandeur, the memory of the honoured dead. We cannot doubt that when the faithful poured forth their prayers at the sacred *confessions* which enshrined in the churches the martyrs' bones, they felt that the majestic sweep of the apse and triumphal arch above them was in a sense the martyrs' monument. Yet it is dangerous to adopt a somewhat far-fetched derivation, when a simpler account of the matter is available. The memorial cella may certainly, to the mind of the early Christian worshippers, have lived on in the apsidal end of the great basilicas, but there is no question that had no memorial cella of this architectural form ever existed, the apse would still have been the dominant feature of the early Christian church. Such a termination to the building was an architectural necessity, and followed naturally from the uses to which it was put.

The apse, that is to say, was *of the essence* of the Christian church, just as the continuous colonnade was the natural feature of the interior of the pagan basilica. For it is here that the difference between the two buildings really resides. The apse was, it is true, an exceedingly common form in Roman architecture, and would have been in place in any building of the times. Apses enshrined the statues in Hadrian's great double temple of Venus and Roma beside the Colosseum; they opened out of the apartments and colonnades in the Thermæ; they adorned the private halls in the palaces of the Cæsars, and, as places for the presiding officials, they terminated the small buildings already spoken of as *curiæ* or *scholæ*. No apse, however, occurs either in the basilica at Pompeii or in the Basilica Julia of the Roman Forum, nor is there mention of it in the account of the normal basilica given by Vitruvius.[1] At one end of the great basilica Ulpia, there was certainly an apse, but its position there, as shown on the Capitoline plan (Fig. 17, p. 91), is very significant. Whether or not it actually formed part of the building, for this is extremely doubtful, it was at any rate cut completely off from the nave by the triple colonnade and gallery, which occupied the narrow end as well as the sides of the basilica. Now this was quite in character with such an edifice, which was, as we have seen, a place in which opportunity was offered for the transaction of several kinds of business at once. Magistrates held their courts there, but the merchants met there also and discussed their affairs at the same time, just as if it had been the open forum. Vitruvius expressly says of a basilica erected by himself, that the tribunal was put in a place apart to avoid interfering with the transaction of business.[2] In other

---

[1] *De Arch.* v. 1. § 4. De Rossi, *Rom. Sott.* iii. 495, says very truly, 'la basilica classica non è di sua natura absidata.'
[2] *De Arch.* v. 1, § 8.

words, *the basilica was not a place devoted to any one single object to which the attention of all within it should be directed,* and it is here that we find its characteristic difference from the Christian church. There, at the time of service, everything depended on the concentration of the attention of all alike upon what was going on about the altar, and it was essential so to arrange the interior that from every point the apse should be seen to be its natural termination. In the basilica there was no such necessity for focussing the architectural effect by directing the eye to one particular part. A curious story in Plutarch[1] illustrates this difference. The tribunes of the people held their court in the basilica Porcia. They complained that access to their tribunal was hindered by the position of one of the interior columns, which they accordingly proposed to have removed. If, as seems probable, this was the central column at the end of the building, the effect of the removal would have been to make their tribunal the most conspicuous object in the basilica, which they would then appear from their seat of judgment to dominate. The attempt was defeated by the opposition of the younger Cato, but it is interesting as a forecast of the change which would have to be made if the business of the basilica was to be concentrated upon a single spot. It was so concentrated when the building was constructed for a Christian meeting-place, and accordingly, in the church of the fourth century the end colonnade is absent, and the apse rules the whole.

A comparison of Figs. 18 and 19 will make clear wherein consists the essential point of difference between the Pagan and the Christian basilica. Fig. 18, which is simplified from Canina's proposed restoration of Trajan's Basilica Ulpia, gives an example of the former. We see here the colonnade and gallery surrounding on all sides the interior space, while

[1] *Cato Minor,* § 5.

a row of windows above admits the light. The roof shows the square cassettes, which seem to have been universal in

Fig. 18.—A Pagan Basilica.

classical ceilings. In Fig. 19 we have a similar building, modified to suit the requirements of a Christian congregation. The end colonnade has yielded place to the arch of triumph with the apse beyond, which at once gives the

FIG. 19.—A Christian Basilica.

structure its character. The gallery is replaced by the wall, which affords space for pictures. The roof shows the open beams, which we still find in the Christian basilicas of Rome and Ravenna. The altar, with the bishop's chair behind it, appears in the apse.

The point here dwelt upon may be finally illustrated by a reference to a form of secular building which seems at first sight to resemble a Christian church more closely than any of those we have passed in review. This is the *private basilica* which existed in the mansions of the great Roman nobles, and especially in the imperial palaces. Vitruvius, in speaking of these mansions, mentions their spacious peristyles, their gardens, their libraries, their picture galleries, *and their basilicas*, where audience was given to clients and causes decided.[1] The arrangement of the Imperial palaces of Rome can now only be traced in the remains of that erected by Domitian on the Palatine, in which we find a general plan similar to that of the normal private mansion already described, carried out with the utmost magnificence in the proportions and decoration of the apartments. Plutarch, when he wishes to give an example of the utmost architectural splendour, mentions the porticoes, the women's apartments, and the basilica in the palace of Domitian,[2] and it happens that the only private basilica of which any remains have come down to us is this very Hall of Judgment of Domitian.

The plan of this (Fig. 20) differs from that of the ordinary festal hall or *œcus* shown in Fig. 3, p. 43, in precisely the same way as the Christian differs from the pagan basilica. The end, that is to say, is occupied by a semi-circular tribunal (not necessarily an apse) dominating the whole, and forming the seat of judgment. The ordinary œcus was a hall of entertainment where the host moved about among his guests, or sat with them at meat; the private basilica was the place where the prince stationed himself to receive

---

[1] *De Arch.* v. 2. [2] *Poplicola*, § 15.

his clients and judge their causes. Hence the same consideration operated in the case both of the Christian church and the hall of audience,—the necessity for focussing the interior by means of one all-dominating feature, and just as the crowds that waited upon the nod of Cæsar, when they entered his hall of judgment, turned their eyes at once to the tribunal where he sat enthroned, so the 'innumerable multitudes' which thronged the churches were conscious, from the first moment they entered the building, that the altar and those about it formed the one centre of interest for all.

So striking, indeed, is the resemblance of the plan of Domitian's basilica to that of the normal Christian church, that some have held that it was actually derived from such private halls without the intervention of the public basilica at all. But this again is a somewhat far-fetched derivation. These halls of audience cannot have been common enough, or sufficiently used by the Christians, to have directly influenced the form of their churches, and it is better to regard the two buildings as having received their form independently, though under conditions which have a striking similarity.

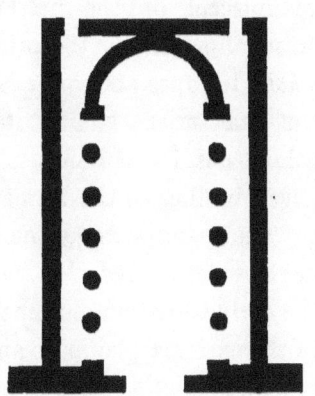

FIG. 20.—Private Basilica in Palace of Domitian at Rome.

The whole question of the origin of the Christian church, so far as it has been treated of in the preceding pages, may be summed up briefly as follows:—The Christians met first in private halls, and when they erected buildings for themselves, these took the form of unpretending lodge-rooms or *scholæ*; they also assembled on occasions in or before the *cellæ* of the cemeteries. At the end of the third and in the fourth century larger buildings were needed, and side aisles were added to the simple halls, which were now lighted in the basilican fashion. Partly as a reminiscence of the *exedræ* of the cemeteries, but chiefly as a natural consequence of the uses to which these buildings were put, they received universally an imposing apsidal termination, which gave them a marked architectural character. Accordingly there is produced from a union of all these elements THE CHURCH OF THE FOURTH CENTURY, with its fore-court and fountain reminiscent of the private house, its oblong plan and tribunal or seat for the presidents derived from the primitive *schola*, its apse and *confessio* recalling the memorial *cella* of the cemeteries; but in its size and grandeur, its interior colonnades, its roof and its system of lighting, a copy of the pagan basilica of the Roman cities.

The materials and the decoration of the basilicas must next be considered. It has frequently been remarked how the early Christian churches differ from the classical temples in the extreme plainness and even poverty of their exteriors, and the existing monuments of Rome and Ravenna are adduced in illustration. It must be noted, however, that literary records of the fourth century give a somewhat different impression. We read there of buildings erected in a solid style of hewn stones, and glistening with marble, and of façades adorned, as some of the Roman basilicas are still adorned, with glass mosaics and coloured incrustations. Some words in the description given by Eusebius of the basilica erected by Constantine at the site of the Holy Sepulchre at Jerusalem will serve as an illustration:—'The vast court open to the heavens was paved throughout of shining stones, and surrounded on three sides by far extending porticoes. On the side which faces the rising sun stood the basilica, a work wonderful to behold, of immense elevation, and in length and breadth of vast extent. The interior of it was covered with variegated marbles, while the surface of the walls on the outside was adorned with polished stones fitly joined together so as to produce an appearance in no whit inferior to marble.'[1] But though the exterior of the basilica might in such instances be treated with a feeling for architectural and decorative effect, as a general rule it seems to have been comparatively plain and unpretending, and in the main the contrast undoubtedly holds good between the classical temple with its great exterior beauty, and the Christian church, which was insignificant externally but full of internal magnificence. These internal adornments consisted in elaborate doors of bronze or carved wood ancient examples of which still remain in S. Sabina at Rome; in splendid gilding and colour either on the open beams of the roof or the cassettes of the flat ceiling, which, in the church described by Eusebius, shone 'like a sea of gold;' in rich hangings over the doors or between the columns of the nave, and in golden lamps and ecclesiastical fittings of various kinds; and more especially in pictorial decoration, which was lavished upon every plain space of wall throughout the edifice.

This pictorial decoration, fine examples of which have been preserved, presents us with a kind of art new both in form and spirit, and of the highest interest and beauty. The earliest creations of the Christian pencil are the wall-paintings in the catacombs, which date from the second to the fourth century. They are remarkable for the light and cheerful character of the designs, as well as for the sketchy but facile execution, which speaks of hands trained in the practice of the fashionable wall-painting of the times. There is nothing that is distinctively Christian in the style of the work, while even in the forms and subjects classical feeling is strong throughout, and in the earliest period even predomi-

[1] *De vita Constantini*, iii. 25, 26.

nates. Slight as these works are, there is a simple charm about them which fascinates the beholder, and is especially marked through contrast with their gloomy surroundings. We seem to hear in them the song of the early Church to herself, in the fulness of a hope which could turn to mirth even in the tomb. These garlands of fruit, these classical genii, these quaint slight figures from Christian and even from pagan legend, are the artless expression of the Church's joy in her existence before a thought of the didactic use of pictures had come in. The designs in the earliest Christian mosaics of the fourth and fifth centuries are inspired by an entirely different spirit. In them the Church appears in worldly glory, grave but triumphant, fully conscious of her mission to teach and to command mankind. If the catacomb pictures are the *lyric* expression of Christian art, the mosaics of the basilicas have something of a sustained *epic* grandeur, which answers to the changed position of the Church. Leaving it to a later age to bring out the *dramatic* side of the Christian story, in representations of the sufferings of the saints or the tragedy of the Passion, the artists of the mosaics chose rather to portray in grand but simple forms the heroes of Christianity. It is a significant fact that no representation of the Crucifixion occurs in Christian art during the period of its first bloom, and it is not till the middle ages proper that it becomes a favourite subject for the artist. It is the same with the representations of the saints. It is not their various adventures which attract the early Christian artist, as they attracted the mediæval miniaturist and painter, but the dignity of their presence, as they stand forth triumphant after suffering, lords in heaven and earth.

We find, accordingly, the apse of the Christian basilicas occupied by the colossal figure of Christ in glory, surrounded by saints, as at SS. Cosma e Damiano at Rome (see Fig.

19), or in the midst of angels, or, as in the grandest of all mosaics that in S. Pudentiana at Rome, in the attitude of Teacher addressing the assembled apostles. Along the nave of S. Apollinare Nuovo at Ravenna finely designed single figures of prophets and saints stand in majestic pose between the windows, while below two long processions of saints—the men on the right, the women on the left—are represented as moving along the walls to the altar end of the church where, on the one side, Mary receives the offerings of the procession of women, and on the other, Christ sits on the throne, in the midst of four sublime angels, to await the homage of the men.

The arch of triumph, one of the most conspicuous positions in the church, was always chosen for a grand display of mosaics. One of the most interesting examples is the medallion bust of Christ in St. Paul's at Rome, one of the few relics of the decoration in the original building. In this the effort of the artist to secure solemn majesty in his design has carried him too far, and there is a forbidding expression of ascetic sternness in the features, from which the finest early mosaics are quite free.

Historical pictures are not so characteristic of this form of Christian art as the single heroic figures, but an extant example of an early date is to be found in the scenes from the life of Christ in S. Apollinare Nuovo at Ravenna. The subjects are treated here with the same epic feeling. The scenes of the passion of Christ are omitted altogether, and He appears only as a dignified figure in the character of Teacher, or as exercising miraculous power. Another series from the Old Testament in S. Maria Maggiore at Rome is not so classically simple in style as the noble pictures at Ravenna, but is treated with much picturesque feeling.

Both the single figures and the historical scenes had a distinct didactic purpose. The former presented an imposing spectacle calculated to impress the beholder with the might of the Church, the latter were to instruct the convert or the uneducated believer in the facts of the Christian story, and rouse him to the emulation of great examples. In the writings of Paulinus of Nola from the first years of the fifth century there is an interesting passage which shows the way in which these sacred representations were regarded. It must be remembered that the early Christian churches and their adjacent buildings were not merely used on stated occasions for service. They were, like the Jewish synagogues, places of resort for the faithful, who congregated there at festival-tide, and passed whole days in and about them. Especially was this the case at the feasts of the martyrs, at which vast crowds came together to pass the night before the sacred day in watching. Eating and drinking, singing, and even dancing in unseemly fashion, were incidents of these vigils and feasts, and Paulinus tells us how he covered the church of St. Felix at Nola and its adjuncts with pictures, in the intention of providing the assembled folk with something better to occupy their minds. The pictures were not symbolic or dogmatic, but had a simple educational purpose unconnected with ritual. This is what he says upon the matter.[1] 'You ask me my object in thus adorning the walls with animated figures. Hear then the reason. The assemblies which the fame of St. Felix brings together are well known. The crowd is great. Here are rustic minds, not wanting in faith, but unskilled in letters, and accustomed long to bow themselves to profane rites with their appetite for their god; these coming now as strangers are brought home to Christ through the merits of the saints. See you how they flock in from all the country round. . . . They have left their homes far away, they have despised the frost,' for their warm faith keeps out the cold,

---

[1] Paulinus Nolanus, *Poema de S. Felice natal.* ix. 541 *seq.*

and now in throngs they are filling the hours of the wakeful night with joy, dispelling sleep by mirth, and by candles the shades of darkness. But would that in all their joy they kept the bounds of temperance, nor quaffed the wine-cup within the holy thresholds! To a sober gladness and a decent service no one would wish to set a limit. Nevertheless I pardon the mistake of their untrained spirits, for their simplicity, unconscious of error, falls through its own warmth of affection,—as thinking in its blindness that the saints rejoice when their tombs are reeking with the odour of wine. Wherefore it seemed to us a good work to deck the whole of the house of Felix with sacred pictures, if haply through the sight of them the forms and colours might seize upon the astonished minds of the country folk. Above the designs are placed their titles, so that the written word explains what the hand has portrayed. There, while the whole multitude in turn point out the pictures one to another, or go over them by themselves, they are less quick than before to think of feasting, and feed with their eyes instead of with their lips. In this way, while in wonder at the paintings they forget their hunger, a better habit lays gradual hold on them, and as they read the sacred histories they learn from pious examples how honourable are holy deeds, and how satisfying to thirst is sobriety. Then comes forgetfulness of wine, the cups grow fewer as the day passes in contemplation, and the time devoted to these sights of wonder leaves but few hours to be spent at table.' The passage ends with some examples of lessons to be drawn from supposed pictures of Old Testament scenes.

The pictures described by Paulinus were mural paintings, which we must suppose to have formed a conspicuous feature in the interior of most ecclesiastical buildings of the time. Wherever it was possible, however, the more costly and permanent material of glass mosaic was employed for the realisation of the designs, and it is only in this form that they have come down to us. The artistic style of these mosaics is simple and elevated, the draped forms are reminiscent of the classic, but have a certain stiffness as well as monotony of treatment which is suitable both to the material used and the effect aimed at; the colours are rich and well harmonised, the setting very careful, and the joints fine. The ground in the earliest and best Roman mosaics is a dark blue, finer in effect than the gold which came afterwards into fashion. The borders are varied and jewel-like in their brilliant hues; some of the most effective, like those in the mausoleum of Galla Placidia at Ravenna, imitate magnificent wreaths of fruit and flowers. It is evident that in these mosaics we see the result of the efforts made by Constantine and his successors to improve the artistic work of their times, and the effect of classical models is very apparent in the earliest and finest examples.

The art was one which gradually decayed as the middle ages approached, and it is only its pallid ghost which haunts the domes of the eleventh century church of St. Mark at Venice. The early examples still remaining to us are sufficient to enable us to form some opinion of the wealth of splendid colour and noble forms which must have adorned some of the more richly endowed basilicas of the age of Constantine.

CHAPTER IV.

THE DOMED CHURCH AND BYZANTINE ART.

AMONG the architectural features on which depends the expression of a building, there are none more significant than the tower and the dome. Nothing stamps more definitely upon a building a special æsthetic character than the employment of the elastic soaring spire, or the reposeful all-embracing cupola, and to these features Christian architecture owes a considerable debt. In the West the tower, originating in early Christian times, becomes under the hands of mediæval builders the feature wherein resides especially that romantic aspiring character of Christian architecture which finds its most perfect outcome in Gothic, while the dome is the favourite form of the builders of the Eastern church, and passing from them (and from the Persian Sassanidæ, see *infra*, Appendix) to the Moslem, was carried by their victorious arms over a great part of the nearer East and of India. It is with the origin and early use of the dome that we are at present concerned.

The question before us is one of the most difficult with which the student of ancient architecture has to deal. The history of vault construction generally among the various peoples who practised it in Egypt, in Mesopotamia, in Hellenistic Greece, in Etruria and Rome, and in eastern and western Christendom, has yet to be written, and one of the most interesting chapters in such a monograph would be that dealing with the fortunes of the dome. An extraordinary fact meets us at the outset, the explanation of which is a hitherto unsolved problem. Unlike other architectural motives, the development of which can generally be traced through successive stages up to perfection, the dome appears in absolute completeness in the earliest example of which we have any certain knowledge. This is the Pantheon at Rome, commenced in all probability about thirty years before the Christian era, and the first in time as in beauty of all the architectural monuments of the Rome of the Emperors. It is not pretended that there were no regularly vaulted domes before that of the Pantheon;[1] but we can only conjecture their existence; there are no actual remains. So far as tangible evidence is concerned, the history of the dome previous to the Pantheon is a blank. Created

'As from the stroke of the enchanter's wand,'

this majestic cupola, the widest, the most beautiful, and as M. Viollet le Duc has pointed out,[2] the best constructed and most stable of all the great domes of the world, seems to set at defiance all ordinary rules of architectural development.

'It is to the last years before the Christian era,' writes M. Choisy,[3] 'that must be assigned the appearance of vaults of concrete'—of which the Pantheon is the finest specimen—'among the monuments of the Romans. Long preliminary trials had without doubt prepared for this important innovation, but no sure trace of these can be found, either in the

---

[1] A distinction must of course be drawn between true arches and domes, those in which the principle of vaulting is carried out and a lateral pressure is exerted by the thrust outwards of the materials, and a *pseudo* arch or vault formed by *encorbellment* (*i.e.* the continuous projection of each horizontal course over the one immediately below it), like the cupola of the Treasury of Atreus at Mycenæ, and many Egyptian arches, in which there is no lateral pressure. See Perrot, *L'Égypte*, p. 113. In the text no account is taken of these pseudo vaults and domes.

[2] *Dict. de l'Arch. Française*, art. 'Voute.'

[3] *L'Art de batir chez les Romains*, Paris, 1873, p. 178.

ruins or in literary sources.' The ingenious author of *L'Art
de batir chez les Romains* hints here at a problem, the importance of which he hardly seems to recognise. Where were
these 'long preliminary trials' carried on? In other words,
what was the previous history of the Imperial Roman
vault which appears in its perfection in this earliest known
example? In face of this perplexing question, it has long
been the custom to conjure by the word 'Etruscan.' The
Etruscans practised the use of the arch, therefore the vaults
and domes of Imperial Rome are explained as the outcome
of Italian tradition. But it is a passage of considerable
magnitude from the tunnel of a cloaca or the arch of an
aqueduct or a city gate to a cupola 144 feet in span, constructed with a skill which has excited the wonder of all
succeeding ages. Where are the intervening steps? In the
history of the extension and beautifying of Rome, which
went on during the last centuries of the Republic, we are
continually brought in contact with Greek influence. To
the Etruscans Rome owed her grand system of conduits and
underground channels, with the style of her older temples.
Her roads were her own creation, and she wrote with them
her title to the earth. Every architectural work, however,
which concerned the beauty and convenience of the city,
and above all the whole apparatus of her later magnificence
and luxury, was an importation from Hellas. 'The Greeks
it was,' writes Heinrich Jordan,[1] 'the conquered Greeks, to
whom the conquerors owed their taste for the artistic adornment of their city.' 'For a long time,' remarks Professor
Adler,[2] 'the architecture of the Republic was insignificant,
until the art of Greece, brought over by victorious consuls,
set aside the old Etruscan traditions.' If this be so, are we
not compelled to look also to Greek models for the grand

[1] *Die Kaiserpaläste in Rom*, Berlin, 1868.
[2] *Die Weltstädte in der Baukunst*, Berlin, 1868.

vaulted constructions of the Empire? We return here to the remarks previously made about the architecture of the great Hellenistic cities, as forming a link of connection between the ancient building traditions of the East and those of the Roman Empire with its modern offshoots. It must frankly be confessed that there is little here to aid us but conjecture. The Hellenistic cities have perished, leaving but slight traces, and there is little evidence but that of probability in favour of our supposing that the Roman brick and concrete with marble veneer, the Roman domes and vaults, even the Roman union of the arch with the column and architrave, had their prototypes in cities like Alexandria and Seleukeia on the Tigris.[1]

Of far more importance in connecting the dome with Eastern rather than with purely Italian traditions, is the fact that in some forms, at any rate, it was a common architectural feature in ancient Egypt and Mesopotamia, while the abundant use in later times of the dome by the Persian Sassanidæ, the Byzantines, and the Mohammedans, seems to show that it has some primeval connection with Eastern regions.[2] In dealing with these primitive Eastern domes a certain caution is needful. What we are seeking for is the

[1] That this was so is the opinion of Professor Adler, *Das Pantheon zu Rom*, Berlin, 1871. As regards the dome there is one passage pointing to its use in Alexandria, though this is by no means as clear as might be desired. The passage is in the late Greek rhetorical writer Aphthonius, and seems to describe a dome resting on a drum surrounded by half columns, the whole upborne on four great pillars.[1]

[2] The Buddhist *topes*, solid monuments without interior spaces, are terminated above in a dome-like form, and bear out this hypothesis.

[1] The passage is given in the original, as any translation would necessarily be also a commentary, and could not honestly be offered without justification of the renderings, for which there is no space. The passage relates to the 'Propulæum,' or monumental entry to the citadel (Museum?) at Alexandria. Προπύλαιον . . . καὶ τέτταρες μὲν ἀνέχουσι μέγισται κίονες, . . . ταῖς δὲ δὴ κίοσιν ἐπανέχει τις οἶκος μετρίας προβαλλόμενος κίονας, οἳ χροιὰν μὲν οὐχὶ μίαν παρέχουσι, παραβαλλόμεναι δέ τῇ κατασκευῇ παραπεπήγασι κόσμος. Ὀροφὴ δὲ τῷ οἴκῳ προῆλθεν εἰς κύκλον· παρὰ (περὶ) δὲ τῷ κύκλῳ μέγα τῶν ὄντων ὑπόμνημα πέπηγεν. See *Zeitschrift für Alterthümswissenschaft* for 1839, p. 376

origin of the dome *as a monumental form*. The primitive Eastern dome seems to have been on a very small scale, and to have been used for subordinate purposes only. On early Egyptian monuments we find the representation of rows of buildings covered with dome-like roofs. These are clearly nothing but store-rooms or granaries of modest dimensions. On Assyrian monuments similar structures are shown, and one plate from Layard's *Nineveh*[1] exhibits an interesting group of buildings covered, some with hemispherical, others with egg-shaped domes. The scene is a country one, and the buildings may be village houses, or, as has been suggested, kilns or ovens. They have certainly no monumental character. Similar structures, also on a modest scale, are to be seen nestling up against a city wall in a design on a Phœnician bowl, figured in Helbig's *Das Homerische Epos aus den Denkmälern erälutert*.

It is interesting to find that domes of this simple kind are used to the present day in the same regions in which we find them figured in monuments three or four thousand years old, and a graphic passage in M. Place's *Ninive* describes how he witnessed the construction, by some village women in Mesopotamia, of a small domed oven, eight feet in diameter, through a process which may well date back to the days of Sennacherib.[2] We have therefore the following results. The dome was employed on a small scale for subordinate buildings in ancient Egypt and Mesopotamia, and was probably constructed of crude brick by the same process in use at the present day. The architects of the Hellenistic cities, who were confined in some cases to the

[1] Second Series, pl. 17.
[2] *Ninive et l'Assyrie*, i. 266, cf. Perrot, *L'Égypte*, 112. In the *Thousand and One Nights*, tale of the enchanted Prince, Night 7, there is mentioned a 'round-roofed hut of mud bricks,' which will serve as a mediæval example.

use of brick, *may* have adopted this form, and *may* have carried it out on a monumental scale. The close connection between Roman architecture and Greek renders the hypothesis a natural one, that the dome came to the Romans from the East through the medium of the Hellenistic Greeks. This is, however, by no means a matter of proof, and we have finally the fact, which is the ground-work of the discussion in this chapter, that *the first monumental dome of which we have either remains or clear record, is the cupola erected by the architect of Marcus Agrippa over the drum of the Pantheon some time before the year* 27 B.C.[1]

This extraordinary hiatus in the history of the dome, between the small mud-brick cottages ovens and storehouses of Egypt and Mesopotamia, and the magnificent vault of the Pantheon, leaves it an open question whether it is most correct to consider the dome an Eastern or a Roman form. On this question depends the view we take of Sassanid and Byzantine architecture. Are these the fruit of an old Eastern stock independent altogether of the West, or were they directly inspired by the splendid triumphs of Roman dome construction? The key to this problem may be hidden among the mounds of rubbish which cover the sites of the once opulent and beautiful Hellenistic cities. If it could be proved that the cupola grew in these to monumental proportions, we should be forced to look upon Roman Sassanid and Byzantine domes as successive offshoots

---

[1] It is necessary to insist the more strongly upon this fundamental fact, because the writer of a recent work of importance, M. Dieulafoy, in *L'Art antique de la Perse*, 1884-85, has lately denied it, though upon wholly insufficient grounds. This writer holds that some of the well-known 'Sassanid' domes, belonging in the common view to about the fourth century A.D., were built in the days of Darius and Xerxes. This theory, a very extraordinary one, will be examined in the Appendix. M. Dieulafoy's hypothesis is accepted by Choisy, *L'Art de batir chez les Byzantins*, Paris, 1882.

from the Greco-oriental stem. If, on the contrary, it was the Roman architects themselves, and foremost among them the creator of the Pantheon, who first made the dome a really important motive in architecture, then the cupola is by right Roman, and the Pantheon is truly the mother-dome of the world. Fresh discoveries on Oriental sites may aid us in this perplexity, but till some new facts in the history of dome construction come to light, or till M. Dieulafoy and M. Choisy can bring forward some better proof of the truth of their recent paradox, there is no reason for wishing to deprive the Roman architects of the meed of honour which has been universally assigned to them as the first great dome constructors of the world.

The æsthetic aspects of this question demand a moment's notice. The dome has been spoken of as a singularly *expressive* architectural feature, and there are those to whom it seems especially connected with Oriental ideas. Here are some words by the late Professor Unger of Göttingen, the historian of Byzantine art. 'One may say that there is no other architectonic form which is so suited as this to serve as the expression of the Oriental view of life, no other breathes to so great an extent the spirit of mystical contemplation. . . . Through the cupola the impression of grandeur, sublimity, and vastness is greatly heightened.'[1] If this is true, it may be partly explained by the obvious connection between the form of the cupola, and that of the vault of heaven, through which it produces naturally the impression of the one, the all-embracing.

When M. Renan says that the desert is monotheistic, he is claiming for the East a special power of nourishing these grand unifying conceptions, but it is, on the other hand, to be remembered that these do not belong wholly to the East, and it is this very idea of *unity* which underlies the

[1] Ersch und Gruber, *Encyklopädie*, § I., Th. 84, p. 351.

policy the thought and the constructions of Rome. Virgil expresses the Roman idea when he writes of the varied gifts in art eloquence and science of the Greeks, and then turns to his own country with the words:—

> 'Rule thou the nations, Rome, imperial! thine
> These arts alone. Peace on the peoples bind,
> The vanquished spare, the proud in arms confound.'

Rome strove to make a unity of the whole world of her possessions. She not only conquered and incorporated in her own body-politic the nations, but she united them by her bridges and roads, which abolished natural barriers, and brought distant provinces into connection. Her mighty aqueducts, which traverse the plains in monotonous succession of arches towards the walls of her cities; her amphitheatres, with their endless iteration of pillar and arch, and their unbroken rings of seats—these are fit emblems of her irresistible course, her levelling, all-dominating policy, before which all limitations, all local varieties, were forced to disappear. In this way the empire of Rome seems only the continuation of the great imperial systems of the East. What it offered the world was the essence of these systems, with the addition of all the apparatus of rational politics and law, which was the creation of the West. M. Place, in his account of Assyrian architecture, remarks justly on the points of similarity it bears to that of Rome, while we may bring out the resemblance further by contrasting both with that of Greece.

The Pantheon, though it has been provided with a portico, is, in its style and character, essentially unhellenic. The Greeks delighted always in the appearance of contrast; with their fine sense of proportion—the foundation of their artistic character—they loved to measure one thing against another, and to note the diversity of parts which are held

---

[1] Virg., *Æn.* vi. 852.

together in harmony in their works of art. Hence they set their temple as a monument upon the pedestal of its stylobate, and throughout it opposed part to part,—the upright column to the horizontal entablature, the capital to the shaft, the architrave to the frieze, the ornamented portions to the plain spaces which set them off. In the Pantheon all is uniform; the decorative forms are repeated, not contrasted; the attention is never roused to balance and to measure; the mind is impressed with a vague sense of sublimity rather than excited to a more active artistic delight. Hence an interior like that of the Pantheon—with its simple divisions, its surfaces so sparingly broken, its immense dome brooding equally over all—conveys a sublime idea of unity, which is perfectly expressive of the character of the Romans. From the æsthetic point of view, therefore, though the dome may have originated in other climes, it is not at Rome an exotic, but a genuine national production. The old traditions of dome construction were absorbed by the Roman builders, and given forth again in a perfected form which has every right to be considered their own property. The importance of a clear understanding upon this point will appear as we proceed, and if this subject seems to be dwelt upon at unnecessary length, it may be urged in excuse that it is a common practice to call all domes built after about the sixth century, whether they are in the East or the West, by the inaccurate term 'Byzantine.' It must always be kept in mind that Rome possessed a world-famous cupola several centuries before the first Byzantine dome, and that during those centuries, dome construction had advanced on parallel lines in the West and in the East, so that the middle ages inherited in the West as genuine a tradition in regard to the cupola as any which flourished in the East.

It is no part of the present subject to discuss the

Pantheon, but a hope may be expressed that before long some complete monograph may be composed on this most important of Roman monuments. Every writer on architectural history or on Roman antiquities has an account of the Pantheon, but we are still in doubt as well regarding its history and its destination, as its system of construction, the original features of its interior, and even its correct measurements. In his official report of the recent excavations around and behind the building, Sig. Rodolfo Lanciani deplores the uncertainty in which the building is involved. 'Il Pantheon,' he writes, 'presenta il fenomeno singolarissimo, di essere l' edifizio antico il più intatto, ed al tempo stesso di rimanere inesplicabile sotto parecchi punti di vista, che concernono tanto la massa quanto i particolari.'[1] It illustrates this remark to find that even that all-important measurement, the internal diameter of the dome, is given differently by almost every authority. Thus, reducing the measurements to English feet, we have from Desgodetz[2] 143 feet 6 inches; from Canina,[3] Isabelle,[4] and Viollet le Duc[5] about 144 feet 6 inches; from Fergusson[6] 145 feet 6 inches; while in the latest monograph on the building,[7] the span of the vault is reduced to 139 feet.

With the Pantheon for a starting-point, let us now glance at the employment of the dome during the early centuries

---

[1] *Il Pantheon, etc., prima Relazione a sua Eccellenza il Ministro della Istruzione Pubblica*, p. 4. Sig. Lanciani is of opinion that the Pantheon was never an apartment of the Thermæ of Agrippa, but, from its first origin, a *temple*.

[2] *Les Édifices Antiques de Rome dessinés et mesurés très exactement*, Paris, 1682.

[3] *Edifizi*, ii. tav. 67-74.

[4] *Les Édifices Circulaires et les Domes*, Paris, 1855, pl. 6.

[5] *Dictionnaire*, art. 'Voute.'

[6] *History of Architecture*, i. p. 311.

[7] Ciro Nispi-Landi, *Marco Agrippa e i suoi Tempi, le Terme ed il Pantheon*, Roma, 1883. (An unscientific work.)

of our era. The most constant use of this architectural
form by the Romans was in connection with their great
bathing establishments. Circular buildings covered with
cupolas are included in the normal *Thermæ* described by
Vitruvius,[1] and we find them occupying in the Thermæ of
Caracalla and Diocletian the place held by the Pantheon in
relation to those of M. Agrippa. One important building
of this kind is known by the erroneous name of the 'temple
of Minerva Medica.' Another class of buildings constructed
by Roman architects with domes are tomb-monuments.
For these the circular form was traditional, and well-known
examples are the tombs of Cæcilia Metella and Hadrian.
When the dome came into use, to employ it for the covering
of such structures was a natural step. Circular temples
were also traditional at Rome, and Vitruvius describes them
among the normal forms of his day.[2] The temple of Vesta,
goddess of the hearth, was always round, and it has been
suggested that the origin of it is to be found in the circular
huts of the herdsmen who settled in primitive days upon
the seven hills.[3] Later on, in the time of the Empire, other
deities also were worshipped in round temples, and of one
dedicated to Helios we read that its circular form represented the figure of the sun, the light being admitted, as in
the Pantheon, by a single opening in the roof.[4] This seems
to imply a dome, while another domed temple formed part
of the palace of Diocletian at Spalatro in Dalmatia.[5] Our
concern is not, however, with pagan but with Christian
buildings. In connection with these the most general use
of the dome in the early centuries is for the covering of
baptisteries. These small round or polygonal buildings, with

---

[1] *De Arch.* v. 10. [2] *Ibid.* iv. 8.
[3] De Rossi, *Piante iconografiche e prospettiche di Roma*, p. 3.
[4] Macrob. *Saturnal.* i. 18, § 11.
[5] This is now considered to have been destined for a tomb-monument.

their domical roofs, appear in the mosaic pictures copied in Figs. 12, 13, 14, as regular adjuncts to the Christian meeting-house. When the first baptisteries were built we have no means of knowing;[1] but both their name and form seem borrowed from pagan sources. They remind us at once of the bathing apartments in the Thermæ, and the fact that Pliny, in speaking of the latter, twice uses the word *baptisteria*, seems to point to this derivation. The Christians also adopted the use of the dome in connection with tomb monuments, and perhaps the earliest Christian dome that can now be recognised is that covering the small memorial cella of the third century in the cemetery of S. Callisto, of which the frontispiece gives a restoration.

In the case of early domed buildings, we are in the same position as in that of the meeting-house; nothing of importance exists of earlier date than the age of Constantine. The earliest existing baptistery is that of the Lateran, said to have been erected in its original form under Constantine. Throughout the Roman world round or polygonal baptisteries seem to have been constantly employed from the fourth century onwards, and the Italians have preserved throughout the separate building for baptism, while north of the Alps the practice has generally prevailed of administering the rite in the churches. Among existing Christian tomb-monuments of domical form the first place is taken by the round building known as S. Costanza, on the Via Nomentana near Rome, erected, according to tradition, by

---

[1] The rite of baptism, as we find it in the New Testament and in the *Teaching of the Twelve Apostles* (vii. 4), is independent of any special place. At the beginning of the third century Tertullian (de Coron. 3) speaks of baptism as a formal and somewhat elaborate rite, but does not imply the existence of a separate building. Somewhat later the Canons of S. Hippolytus (xix. 12) seem to regard the place of baptism as distinct from the church, while the discourses of St. Cyril of Jerusalem (middle of fourth century) presuppose independent baptisteries of considerable size.

Constantine for the sepulchre of his daughter about 336 A.D., and adorned with the earliest existing Christian mosaics. The constructive aspects of this building will be noticed on a subsequent page. Some other important domed buildings of this time are no longer preserved to us, and among these the foremost place is due to the round structure erected by Constantine between the years 326 and 334 A.D., on the supposed site of the Holy Sepulchre at Jerusalem.[1] This seems originally to have consisted in a small dome supported on a ring of twelve columns, representing the twelve apostles, and in this form it is represented in an early Christian mosaic in S. Apollinare Nuovo at Ravenna. The influence of this building upon later architecture has been very marked. If the martyrs' graves in general received, as we have seen, such extraordinary reverence, still more would naturally be lavished upon the supposed tomb of our Lord, and it is not surprising to learn that pilgrims who had visited the shrine erected copies of it on their return to their own country. Such a copy exists in the church of S. Sepolcro at Bologna, one of the seven small buildings grouped together under the name of S. Stefano, and it is said to have been erected by S. Petronio about 430 A.D., after a visit to the Holy Land. On the platform of the temple at Jerusalem stands a beautiful and enigmatical building known as the 'Dome of the Rock,' and less correctly as the 'Mosque of Omar.' Whether this be Arabian (Adler), Byzantine of Justinian's time (Sepp), or,

---

[1] Around the buildings of Constantine at the Holy Sepulchre (Euseb. *de Vita Const.* iii. 31), has gathered a mass of controversy, in which is also involved the famous 'Dome of the Rock.' For the different opinions expressed on these subjects see Fergusson, *The Holy Sepulchre and the Temple at Jerusalem*, 1865; Unger, *Die Bauten Constantins am heiligen Grabe*, Göttingen, 1863; Adler, *Der Felsendom und die heilige Grabeskirche zu Jerusalem*, Berlin, 1873; Sepp, *die Felsenkuppel eine Justinianische Sophienkirche*, München, 1882.

as Dr. Fergusson and Professor Unger have maintained, in part at least the original church of the Holy Sepulchre built by Constantine, it is a structure of great historical importance. A comparatively modern development of dome-construction appears here anticipated, and the cupola presents itself for the first time as a prominent *external* feature. The 'Dome of the Rock' has had, too, a curious influence upon western architecture, and has become, through a pardonable blunder, the parent of all the churches of the Order of the Templars. These crusading knights imagined it to be part of the temple of Solomon, and adopted its form for their own buildings, so that wherever in the West there appears a church of the Templars, there we see perpetuated the plan of this so-called Mosque of Omar.

The above is sufficient to illustrate the use of the domed form by the Christians for baptisteries and sepulchral monuments. We have now to consider its employment for the more important purposes of congregational assembly.

The first domed churches built for Christian worship are of special interest, for from them sprang Byzantine architecture, with its various offshoots. So far as our information extends, the builder of the earliest domed church of any magnitude was Constantine,—its locality the famous city of Antioch in Syria, the nurse of the infant Christian community. About the year 327 A.D., Constantine erected here a building, which Eusebius calls 'a kind of church altogether unique for its size and beauty.'[1] Its ground-plan was an octagon, and high above, supported on precious columns, rose the light and lofty cupola, flooded with light which played upon its roof, and won for it the name given to it by St. Jerome of *Dominicum Aureum*, 'the golden House of the Lord.' There seems good ground for believing that this

[1] *De Vita Const.* iii. 50.

building, the great ornament of the 'metropolis of the East,' —so Antioch was called,[1]—which contained at the end of the fourth century a larger Christian community than Constantinople,[2] was the starting-point of the rich development of domical construction in the Eastern empire, and among the adjacent peoples. It is true that we meet here once more the question discussed in the preceding pages, the question whether or not the Hellenised East possessed an independent tradition of this domical construction, which could have produced Byzantine architecture without the intervention of Rome. It is not possible in the present state of knowledge to give a confident answer, but something may be done by calling attention to the few definite facts which form our most important landmarks. This 'golden church' of Antioch occupied part of the site of some Thermæ of the Emperor Philip the Arabian, and its plan may have been adopted from that of the circular bathing apartment of these Thermæ. Of the octagonal form chosen for it we have another example in Constantine's baptistery of the Lateran at Rome, and these two facts would appear to connect it with Roman traditions of building. The words just quoted from Eusebius may be taken to mean that it was designed in a style new to the regions of Syria. It is a noteworthy fact also that among the Christian edifices of central Syria described by de Vogüé, the polygonal and domed form is quite exceptional,—the only two important examples of the style, the churches at Ezra and Bosrah,[3] dating from the beginning of the sixth century,—though there, if anywhere, we should have expected this supposed Eastern tradition to have asserted itself with prominence from the earliest days of Christian architecture. If Constantine builds over the

---

[1] S. Hieron, *Contra Joann. Hieros.* 37.
[2] S. Chrys. *Hom. de Consonantia utr. Test.* I.
[3] De Vogüé, *Syrie Centrale*, etc., pl. 21, 23.

Holy Sepulchre a light cupola, resting on a ring of columns, a new and possibly Hellenistic form,[1] he repeats the form, as we shall see, a few years afterwards at S. Costanza near Rome. The admired church at Antioch, the loss of which we have greatly to regret, was copied in a building erected at Neo-Cæsarea in Asia Minor by the father of St. Gregory Nazianzen, while before the close of the fourth century there must have already existed at Milan in North Italy the original edifice of S. Lorenzo, which proves how far the early Christian builders of the West had advanced at this time in the development of dome construction. The cupola did not take its position as the characteristic form of Eastern Church architecture till some two centuries after Constantine, and even then it was not exclusively Eastern, but remained in use in well-marked, though comparatively unfrequent, examples in the West.

Up to the reign of Justinian (A.D. 527-565), the vast majority of churches erected throughout the Empire were of the oblong basilican type, or, as the Greeks called them, δρομικοί, from the resemblance of their plan to the stadium or race-course. It was not till the first half of the sixth century, when Justinian's architects erected S. Sophia at Constantinople, that the full capabilities of the domed style were made apparent, and this form was definitely adopted by Byzantine architects. From this time onwards almost all the churches of the Eastern Empire exhibit the domed style, which spreads into the neighbouring lands, and produces on the one side the architecture of Armenia, and on the other that of Russia, while it exerts a marked influence on the constructive forms of the Arabs. Western architects continue, however, their independent use of the dome, and the beautiful S. Vitale at Ravenna dates from about the same time as S. Sophia, while at the end of the eighth

[1] *Ante*, p. 83.

century we have the famous minster of Charles the Great at Aachen, the form of which reappears in several churches of the Rhineland. From the eleventh century onwards cupolas are not unfrequent in conjunction with basilican ground-plans. The Duomo of Pisa (enlarged to its present form in 1063) is the earliest important Western example of a dome covering the crossing of a cruciform church, and cupolas in similar positions are frequent features in Romanesque minsters. It has been urged that in these domes we have a direct result of Byzantine influence, but the existence in the West of a tradition of dome construction based upon Roman and early Christian models, renders the modern critic cautious about assuming Byzantine influence in the West, except where there is positive evidence that it was exercised.

That evidence is clearest in the case of St. Mark's at Venice, which is undoubtedly Byzantine in form and decoration. It is not to be denied again at St. Front at Périgueux in Western France, and in other churches of that region which are derived from St. Front, though here it is only the bare form which is Byzantine, the decoration being of purely classical type. S. Antonio at Padua has also a pronounced Byzantine character. At the time of the Renaissance in Italy and Western Europe, domed constructions became again of the first importance, and it is interesting to note that the Pantheon becomes once more, in the hands of Brunelleschi, the starting-point of a new development, which culminated in the cupolas of St. Peter's at Rome, and St. Paul's in London.

From this brief sketch of the history of the domed church we now pass on to consider its development in its more strictly constructive aspects. The characteristic difference between the Christian basilica and the domed edifice resides here. The former was an admirably convenient building for its purpose, but as will be seen in the next chapter, it was structurally crude and incomplete. The latter was, on the contrary, architecturally perfect in its earliest form, but not suitable for the purposes of a Christian meeting-house. The development of the two buildings proceeded therefore on opposite lines; the loosely-constructed basilica became the severe and compact Romanesque minster, while the too simple and uniform round building becomes broken up in such a way as to provide the requisite architectural divisions and a commanding position for the altar. The Pantheon resembled the pagan basilica noticed in the last chapter, in that it was nowhere focussed, and had no natural station for the altar, nor any divisions which might serve to keep the classes of the congregation separate. Nor again did its severely regular form admit of those side-buildings so necessary for vestries and the like, while its circular plan rendered it difficult to connect with other buildings. The development of this form of edifice will, as we shall see, proceed in the direction of giving to it the convenient accommodation offered by the Christian basilica, while its architectural character and the fine effect of the dome are still preserved.

The form of the Pantheon is too well known to need description. Its construction is undoubtedly very scientific and is not yet thoroughly understood, but its general plan is of the simplest. A circular drum half as high as its diameter is completely covered in by a hemispherical dome, fitting evenly upon it in every part, light being admitted through a single opening at the summit of the vault. The first step towards a freer treatment of the ground-plan is to be observed in the building known as the temple of Minerva Medica.[1] There is here also a drum and a hemispherical cupola surmounting it, but the severe simplicity of the Pantheon is abandoned. The height of the building is no longer equal to its diameter, but exceeds it, and the form of the ground-

[1] Isabelle, *Les Édifices Circulaires et les Domes*, Paris, 1855.

plan is no longer circular, but polygonal. This leads at once to a modification of the form of the dome. We have no longer a cylinder for its support, but a polygonal drum of ten sides. On this is placed a hemispherical dome like that of the Pantheon, but, as will easily be seen, it will no longer exactly fit. We may either construct it of such a size as to rest upon the extreme corners of the polygon, or else make it a little smaller, so as to be supported upon the centre of each side, the corners projecting beyond. In both cases some more work would be required to complete the construction. This would be obviated if the dome, instead of being round, were itself polygonal, that is, were composed of triangular slips meeting in a point above, each one of which fitted exactly on to the top of one of the sides of the base. This is a form of dome of which we shall find examples, but it is the hemispherical cupola with which we are at present concerned. In the temple of Minerva Medica the dome rests on the corners of the polygon, and would naturally project a little beyond the flat sides. Let the reader imagine that he has before him a model of the form represented in Fig. 21, where the side A is shown as overlapped by the domical covering. If we suppose a knife passed up along the sides, so as to cut off the projecting parts of the dome, we should find that above each side there would be left a round arched aperture, as shown in the drawing over B. The wall may then be continued upwards to fill in this aperture, as at C, and if this were carried out all round, we should again have a complete construction like the Pantheon, though with this important difference on which much of the effect of the interior would depend,—we should miss the effect of the full hemispherical dome, for the vault would seem to begin, not at the points where it rests on the corners of the polygon, but only at the level, D, where its surface rises clear above the upper portion of the sides, which, as we have seen, cut into it.

The cupola would accordingly have a flat saucer-like appearance, and the grand sweep of the complete hemisphere would be lost. The effect of such a cupola is to be seen in the Baptistery of Ravenna, where the dome is joined to an octagonal drum in the way just described.

This flatness of effect would be obviated if the dome, instead of touching the corners of the polygon, rested on the middle of the sides, as shown in Fig. 22. The construction

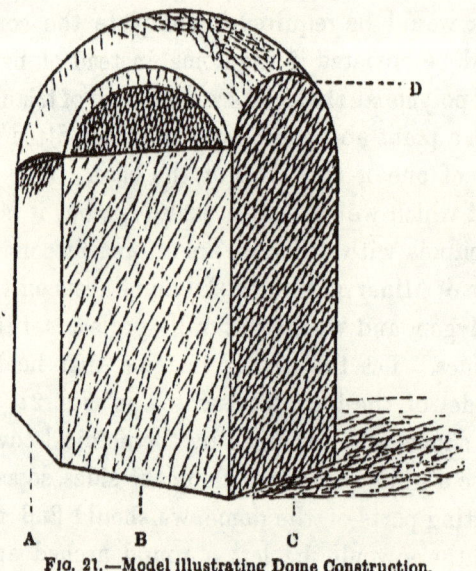

FIG. 21.—Model illustrating Dome Construction.

would have to be completed in the interior by building upwards and outwards from the corners to meet the portions of the edge of the cupola, A, which would be hanging unsupported, while on the exterior the projecting corners of the polygon, B, would require to be masked. The result would be at first sight the same as before, but there would be the important difference that the portions which fill up the corners, whether they are marked off from the dome or merged into it, would be *something extra added to the dome*, instead of being subtracted from its surface, so that the full extent of the hemisphere would be displayed, free of all interference. It is in the treatment of these corners and their connection with the dome that the great constructive advances were made which culminated in S. Sophia, and to which we shall accordingly have to return.

With regard to the question of support, the dome of the Pantheon is to outward appearance borne up equally at every point by the circular drum, the upper part of which, above

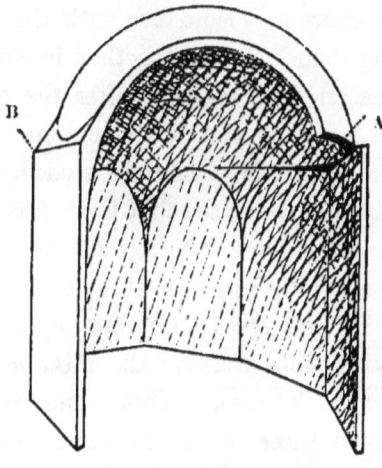

FIG. 22.—Model illustrating Dome Construction.

the niches on the ground-floor, seems a solid mass. The appearance is, however, deceptive. The dome is not homogeneous in structure, for it is in all probability formed of ribs of brick running up to meet in a solid ring surrounding the aperture, the spaces between the ribs where the familiar cassettes appear being formed of nodules or *scoriæ* imbedded in cement. The walls again do not form a solid cylinder, but are scientifically put together with numerous discharging arches, and contain in their thickness chambers and galleries placed, it may be assumed, in order to correspond with the variation between the brick ribs and the concrete filling-in. It is calculated by Prof. Adler that there is actually employed not much more than half the material which would have been necessary for the construction of a solid drum.[1] There is, therefore, even in the Pantheon, an effort to gather the pressure into certain points, meeting it there by special supports, and this proves that there is only partial truth in the idea that the Romans resisted pressure merely by brute mass, not by calculated supports. It is, at the same time, true that it was more in accordance with the genius of the Romans to bring their material together in abundant quantity than to treat it with economy after the manner of the Gothic builders, who concentrate attention only on those points where resistance is absolutely needed. At Minerva Medica we find, perhaps for the first time, this concentration clearly marked. The dome is built with ribs of brick resting upon the corners of the base, and these corners are fortified on the exterior by buttresses counteracting the thrust of the ribs, and relieving from pressure the intermediate portions which rest upon the flat sides. These sides are built thinner in proportion than those of the Pantheon, and are pierced with windows admitting light to the edifice without the necessity of an aperture in the roof. Below the windows the walls are built out into apses which form projections on the exterior, and which, in themselves natural and obvious features in Roman buildings, are of structural advantage

---

[1] *Das Pantheon zu Rom.*, Berlin, 1871. Prof. Adler's bold restoration of the Pantheon, by which the obnoxious attic story in the interior is abolished, and the grand niches opened out to their full height, deserves the warmest recognition on the part of all lovers of this unique monument. It is right, however, to add that this idea is not a wholly new one, since it is noticed by Canina, who decides against it. Sig. Lanciani (*Prima Relazione, etc.*, p. 13) admits the merit of the idea, but is not prepared to accept such a restoration.

here in giving strength to the wall. They act really as hollow buttresses resting against the wall and giving it lateral support, so as to counteract the thrust of the vault.

Another Roman edifice presents us with an important advance in the direction of dividing the inner space and preparing the building for congregational purposes. This is the already mentioned S. Costanza, of which Fig. 23 gives the ground-plan. The novelty here consists in the fact that we have no longer a single drum with its cupola covering the whole interior, but a central space, A, roofed with a dome, and around this a covered passage, B B, which is to the centre just what the side aisles of the basilica are to the nave. The similarity is carried still further by the central drum being pierced with windows above the roof of the aisle, after the fashion of a basilica.[1] An equally important feature is the roofing of the passage or aisle with a continuous tunnel vault, which, with the massive

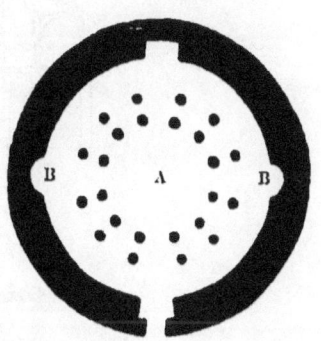

FIG. 23.—Ground-plan of S. Costanza.

walls flanking it on the outside, forms the lateral support of the central cupola. The direct support is given by a double ring of columns, to which, as in the basilica, is given the apparently unsuitable task of bearing up the unbroken mass above. A further and most important stage in the development of the domed church is illustrated in S. Lorenzo at Milan, one of the most beautiful and interesting monuments of early Christian architecture. A ground-plan, with corresponding section, is given in Fig. 24. Here again we find a central

---

[1] Fig. 14, p. 75, shows a baptistery constructed with central space and passage round, while the earlier designs, Figs. 12 and 13, have the simple form of the Pantheon.

Section.

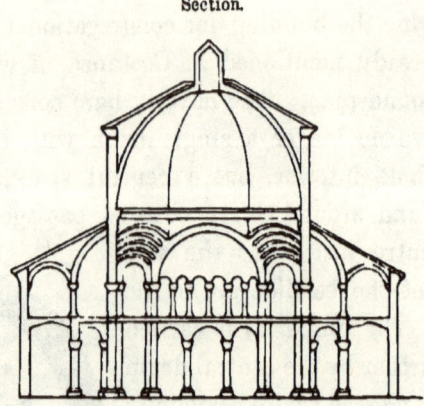

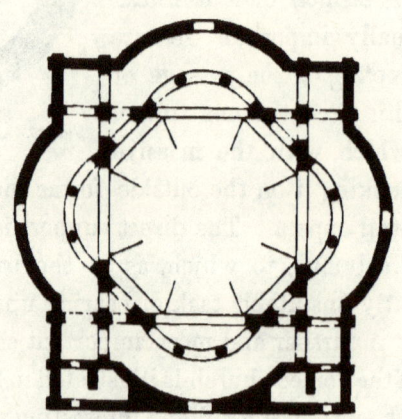

FIG. 24.—Ground-plan of S. Lorenzo at Milan.

cupola with surrounding spaces. These do not, however, merely repeat the form of the central space as at S. Costanza, but are planned so as to make on the outside a square, a form which is far more convenient for connection with other buildings than the circle or the polygon. The central drum is in this building octagonal, and is upborne, not by columns, but by solid pillars. Lateral support is given to it by apses on four sides of the octagon, acting as at Minerva Medica after the manner of buttresses to counteract the outward thrust of the cupola. The other sides of the octagon are backed by additional pillars. Around the central space are the aisles filling up the square formed by the external walls. These walls are, however, thrown out again in shallow apses, so that the square form is somewhat masked. The aisles are two-storied—that is, they exhibit the basilican feature of a gallery opening into the central space under the dome. It is to be noted that the cupola is not round but octagonal, with eight triangular vaulting fields corresponding to the eight sides of the drum from which it rises. In adopting this form the architect has made a sacrifice for the sake of ease of construction, for there can be no question of the superiority in effect of the unbroken hemisphere, and we note that in S. Vitale at Ravenna, which resembles S. Lorenzo, the circular form has been retained.[1]

[1] S. Lorenzo, which is far less often visited than it deserves, was probably erected at the time of the great prosperity of Milan, under the headship of St. Ambrose, in the latter part of the fourth century. In the sixteenth century the cupola, with other portions of the building, fell into ruin, and was restored with only slight alterations by the architect Martino Bassi, who has left an exact account of the old state of the building and of his restoration. 'Il Tempio Vecchio di San Lorenzo,' he writes, 'che rovinò era dell' istessa forma che si va rifacendo,' etc. It is certain, therefore, that we have in it an early Christian monument, and one of the most important of its class. The report of Martino Bassi is to be found in a work entitled, *Dispareri in materia d'architettura*, etc., Bressa, 1572.

These examples are sufficient to show that on Italian soil, and within the limits of the fourth century, some very important steps had been taken towards evolving a church suited to the requirements of a Christian congregation out of the simple structure of the Pantheon. It is surprising to see, moreover, how in many points of construction which we admire in the fully developed churches of the middle ages, the Romanesque and Gothic builders had been preceded by the Roman architects of early Christian times. The structure of Minerva Medica shows us the concentration of pressure upon fixed points of support, which is such a feature in Gothic. The use of pillars instead of columns for the support of the walls and roof of the central space was established at S. Lorenzo. The employment of the vault of the side aisles to sustain laterally the main vault,—a marked characteristic of the Romanesque churches of the South of France,—occurs in Constantine's edifice of S. Costanza. The one point in which the basilica still had the advantage over the already far more architecturally developed domed churches was the position of the altar. At S. Lorenzo the altar seems to have been placed in the centre under the dome, but this was not nearly so advantageous as its position in the grand apse which formed the termination of the nave of the basilica. The question now was, whether the advantage possessed in this respect by the basilica could be combined with the science and beauty of the domed style, and this question, with others relating to the construction of cupolas, was finally answered, not on Italian soil, but at Byzantium, in Justinian's church of S. Sophia.

The general form of this famous edifice will be understood by a reference to Fig. 25. It sums up within itself all the points of interest connected with Christian domed architecture, and contains the solution of all the problems which occupied the builders of the previous age. It is also a

OF THE DOMED CHURCH IN S. SOPHIA.

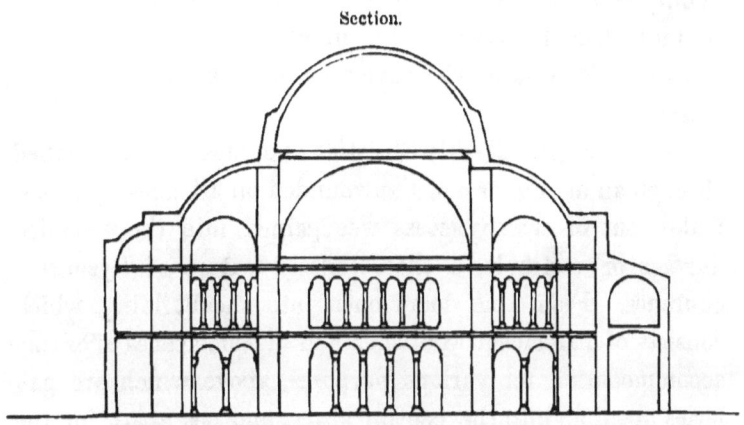

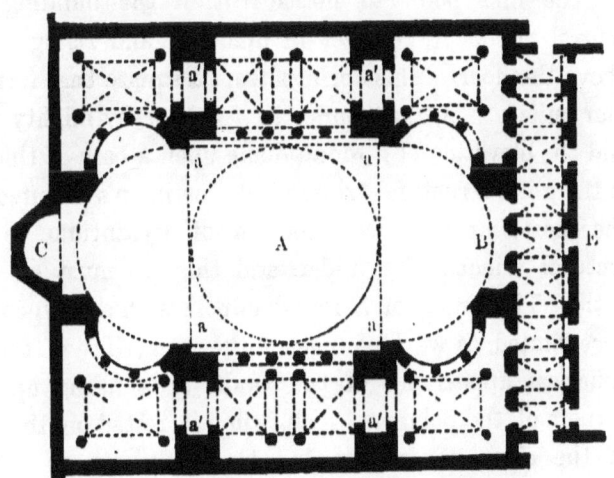

A Central cupola.
a a Supporting pillars.
a' a' Supplementary pillars.
B Entrance.
C Terminal apse.
E Atrium.

Fig. 25.—Ground-plan of S. Sophia at Constantinople.

building of the rarest beauty, the interior effect of which, for form and decoration, has probably never been equalled. It is only possible here to touch upon a few of the more important characteristics of this unique monument, while it must be left to those who have seen it to describe its artistic charm.

S. Sophia, like all early Christian churches, was approached through an atrium or court surrounded on all sides by colonnades, out of which access was gained into the so-called *narthex* or vestibule, a characteristic feature of Byzantine churches. From this doors open into the building, which consists of a free central oblong, and of side-spaces affording accommodation for various purposes, above which are galleries opening into the central space, and set apart for the use of women.

The first point to notice within the building is the central cupola A, 107 feet in diameter, and rising 180 feet above the floor. This springs from a square base formed by four pillars, a, a, a, a, connected together by mighty arches, and the elevation of a round dome upon a base of this shape is the great structural triumph of Justinian's architects, and the most important contribution of Byzantium to architectural science. To understand this we must go back a little. The union of a round cupola with a polygonal base necessitated, as we have seen, either a sacrifice of the hemispherical appearance of the vault, or a filling up of the corners of the polygon which, when it rested on the middle of the sides, its edge failed to touch. At S. Vitale, at Ravenna, begun in the year 526, these corners are filled in by small niches, which have neither connection with each other, nor structural significance. An almost contemporary building at Constantinople, SS. Sergius and Bacchus, erected at the opening of Justinian's reign, shows an important advance in this respect. The fillings up of the octagon are here brought into organic connection with each other and with the dome. They are so formed as to be all sections of one and the same hemisphere, and their upper corners all meet so as to form a complete ring, which is marked with a projecting cornice encircling the edifice. In this cornice we find again what was afforded in the Pantheon, a continuous circular base for the cupola, the construction of which begins from this point. It is as if the builder had started with a cupola in the form of that at Minerva Medica, resting on the corners of the polygon, and cut into by the upper portions of the sides, and had stopped it as soon as it was free of these, and surmounted it with a cornice, on which he erected a fresh cupola in its full hemispherical form. These lower structures are called by architects *spherical pendentives*. When they are employed, the number of the sides of the base is immaterial; they can be made to fill in the corners of a square as well as of an octagon, and it was their introduction which enabled Anthemius of Tralles, the designer of S. Sophia, to bear up his dome on supports so few and so slender that it seems, as Procopius said of it, to be 'floating in the air,' or to be 'suspended by a golden chain from heaven.'[1]

This then is the chief point of structural interest in S. Sophia. On returning now to the section and ground-plan in Fig. 25, and comparing them with those of S. Costanza and of S. Lorenzo, we notice that while in the latter the central space is followed regularly all round by the aisle or passage, in S. Sophia two sides of the central square are thrown completely open, so that a vast hall extends from B to C, through the whole length of the edifice. This immense free space, 250 by 100 feet, and 180 high to the top of the central dome, answers, as will readily be seen, to the nave of a basilica, and contains, like it, the altar in an apse at the

[1] Procopius, *De Ædificiis*, i. 1.

extreme end at C, while the lateral spaces entered through colonnades on the other two sides of the central square, correspond to the lateral aisles of a basilican church. We have thus in S. Sophia a bold attempt to unite *a basilican ground-plan* with the domical construction which, in the earlier churches, had been connected with ground-plans of a form corresponding more or less exactly to that of the cupola above.

With regard to the lateral support of the cupola, this is gained at the sides by supplementary pillars, a' a', and by the solid vaults of the side aisles and galleries. In the direction of the line from B to C, on the other hand, support is gained by a most brilliant extension of the old plan of using an apse as a buttress, which we find employed in earlier buildings. Two grand semi-domes of the same diameter as the central cupola are introduced on each side of it, and again receive lateral support from apses at their sides, as well as from the apse at the altar end at C, and from the vault of the gallery over the entrance-vestibule. In this way the whole building is scientifically built up, one part supporting the other, and an organic connection binds together the whole.

The beauty of the interior effect of S. Sophia evidently depends in great part upon the way in which the eye is led up from the lower to the higher vaults, till it finally rests within the great central cupola. The following words from an architect who had exceptional advantages for the study of the building, lay special stress on this point. 'The total impression which this building, with its many connected parts, makes upon those who enter it, is one of grandeur, sublimity, and splendour. The way the spaces unfold themselves before us is overpowering. At first the eye glances quickly over the wide nave, penetrates deep into the side-halls and the women's gallery, and then lifts itself from arch to arch continually upwards into the soaring dome, the central space at the summit of which, over 30 feet across, is visible even from the lintel of the door by which we enter. Every step that we take forwards opens out new views on either side, and the wealth of gorgeous material, combined with the harmony of the proportions, awakens in the mind of the beholder the feeling of satisfaction and content.'[1]

The dome of S. Sophia appears, from the first moment of entrance, to rule the whole interior, and embrace all that the edifice contains within its unbroken sweep. By this it emphasises in the strongest way the centre of the building, and the question naturally occurs how it is possible for the effect of the long nave to be obtained, without the introduction of an opposing element. Some indeed may doubt whether the builders of S. Sophia have really succeeded in their bold attempt to unite the advantages of the basilican and the domed constructions. The open view along the nave to the apse at the altar end is, as we have seen, the great feature of the basilica, and it may be argued that the attraction of the eye towards the end of the nave in S. Sophia must interfere with its appreciation of the special æsthetic effect of the round cupola, which seems designed to keep everything within itself.

It is clear that this difficulty was present to the minds of the architects, and that they provided against it by their bold and original plan, by which the lines of the circular dome are carried down through the oval composed of the two half-domes connected with it, into the longitudinal form of the nave. The half-domes are thus as important from an æsthetic as from a constructive standpoint, and their employment by Anthemius is one of the most brilliant achievements of architectural genius. Through their interposition

[1] Salzenberg, *Altchristliche Baudenkmale von Constantinopel*, Berlin, 1854, p. 17.

the 'principle of centralisation is carried fully out, but, at the same time, modified,' and the building becomes as much an artistic as it is a structural unity.

The interior effect of S. Sophia was enhanced by the abundant light which played through it, and must have brought out into clearest relief its exquisite colouring and ornamentation. With regard to this, Procopius tells us that 'one might believe that the inner space is not lighted from without by the sun at all, but that radiance dwells actually within it, so vast a flood of light pours itself abroad throughout the shrine.'[1] 'For the lighting of the interior,' writes Salzenberg, 'there is no word but brilliant; a flood of light pours itself through the house of God. The East sends its first rays through the six large apse windows into the nave, and the evening sunshine glowing through the large western window, bathes the vault in fire.'[2]

The decoration in form and colour, upon which these sunbeams fell, was in the finest Byzantine style, that style which plays so large, though so ill-understood, a part in the artistic history of Christendom. It is not the purpose of the present essay to discuss the art of Byzantium, but we cannot turn away from Justinian's masterpiece of architecture without a word on the decorative arts, which reached in his reign their culmination, and the highest achievements of which were called forth for its adornment.

Byzantine culture was spoken of on a previous page as a mixture of Roman, Hellenic, and Oriental. Constantinople was a Roman city, and had been founded as the chief seat of Roman power; it was at the same time only the most splendid and important among several capitals which had divided between themselves the duty of providing for the Imperial administration, undertaken formerly by Rome alone. Before the time of Constantine, Trier, Milan, Nicomedeia,

[1] Procopius, *l.c.* i. 1. [2] Salzenberg, *l.c.* p. 25.

and after him, Constantinople and Ravenna, were all centres of government. Each of these cities was a seat of the same Roman rule, and when any one of them ceased to bear sway, the rest were responsible for its provinces. Thus, when Italy and the West had passed for a time under the power of the barbarians, Justinian, as the representative of Rome, issued forth from the unconquered capital of the Eastern provinces, to attempt the work of its reconquest. It would have been, as Mr. Freeman remarks, a wiser policy to abandon the West to the Teutonic peoples, and devote all the power of the Empire to consolidating its possessions in the East.[1] But Justinian's position as Roman emperor seemed to demand the practical assertion of his authority over the West, as well as over the East. This is only one example of the extent to which the Roman tradition lived on unbroken at Constantinople. Her inhabitants considered themselves Romans, and Latin remained till about the ninth century the official language of the court, the government, and the law.

But, secondly, Byzantium was originally a Greek city, and the ancient seats of Greek civilisation lay all about her. In everything that was not official she was strongly Hellenic. Especially was this Greek influence strong in art. Constantine had, as we have already noticed, collected in his new capital a crowd of the finest works of the Grecian chisel from the various Hellenic cities,[2] and even his original church of S. Sophia contained a collection of antique statues. The way was thoroughly prepared for a long survival at Constantinople of Hellenic traditions of form.

Thirdly, the position of Constantinople on the borders of Asia, and her special responsibility for the Oriental pro-

---
[1] Freeman, *Historical Geography of Europe*, p. 107.
[2] St. Jerome says, 'Constantinopolis dedicatur, pene omnium urbium nuditate.'—*Chron.*, ad ann. 334.

vinces, brought her into constant connection with the East. First with the Persian court, and later, with that of the Arabian Caliphs, the rulers of Byzantium held close political intercourse, while through her markets was incessantly flowing the stream of commerce from Asia to the West. The relation of Byzantium to the East was one of interchange. She received much, but had also much to give. As the guardian of the treasures of classical culture, she was in a position to impose upon the less civilised peoples who had already been dazzled by the glory of Alexander and the Seleucidæ, while the East, as the primeval home of luxury, could offer gifts of ever increasing value to the degenerate descendants of the Roman and Greek citizens.

It would be difficult exactly to distinguish the Latin, Greek, and Eastern elements in Byzantine art, and indeed the preliminary distinction between Hellenic and Roman is by no means easy to draw. There is no question, however, that the fundamental form of Byzantine architecture—the dome—was, proximately at any rate, Roman, though to the Greek builders employed by Justinian must in all probability be ascribed the invention of the spherical pendentive, by which dome construction was brought to perfection. In the case of the arts of form, all that was essential was Greek. The Greeks had taught the Romans the treatment of the human figure and drapery, and, as far as they could receive them, the laws of taste and of composition. The tradition thus established was carried on at Byzantium, and a certain artistic culture and *savoir faire* characterises the Byzantine artist from first to last, so that however low he may fall, there is none of that rudeness, that 'groping in the dark' about his work, which mark the productions of the mediæval workman of the West, and proclaim his want of a sound tradition. Sculpture in the round was but little practised at Byzantium, and this, the most important of classical arts, the crown of the achievement of the Greeks, was the first to perish. It is true that the last great plastic work of antique art was set up at Byzantium in the colossal equestrian statue of Justinian, in heroic guise, which stood upon a column 100 feet high near S. Sophia; but this, we learn, was executed by a Roman statuary. Ecclesiastical feeling was from the first unfavourable to statues in sacred edifices, and early in the eighth century the iconoclastic movement put a stop to their production altogether. The technique of bronze casting was, however, still preserved, and was reintroduced from Byzantium to the West in the mediæval period. Sculpture in relief, especially on ivory, representing both figure-subjects and ornamental foliage, flourished through the whole period of the bloom of Byzantine art, which lasts to the taking of Constantinople by the Latins in 1204. Greek feeling nowhere survives more freshly than in the delicate Byzantine ivory carvings, of which Labarte reproduces some exquisite examples.[1] Of the sculptured forms of decoration applied by the Greeks to their architecture, the acanthus, with other foliage that presents the same sharply accented forms, was especially delighted in by the Byzantines, but they treated it with a certain amount of Eastern feeling, which will be noticed later on.

In painting, which became the Byzantine art *par excellence*, technique and drawing rested on classical tradition, though the subjects soon became almost exclusively ecclesiastical.

Some special forms of decorative art were largely practised at Byzantium. One was incrustation in coloured marbles, arranged in panels or geometrical patterns for the clothing of the brick and rubble walls of buildings. This

---

[1] Labarte in his *Histoire des Arts Industriels* gives a beautiful collection of illustrations of the masterpieces of the decorative arts of Byzantium, with excellent criticisms on them.

style of work originated in the splendid Greek cities of the Hellenistic period, as the name of one kind of it, *opus Alexandrinum*, clearly shows. It was in constant use in the buildings of imperial Rome, and its introduction into Byzantium may either be ascribed directly to Roman influence, or may be looked on as a survival from Hellenistic times. Another form was glass mosaic, which the Romans of the Empire had sparingly practised, but which became from the time of Constantine the characteristic form of decoration for Christian churches. Byzantine mosaics differ from those of Italy in the very large employment of gold. The art of gilding a vitreous cube with gold leaf, which is fixed by melting over it a transparent film of glass, seems to have been a new technical process invented by the Byzantines, and these cubes are used everywhere as the ground of Christian mosaics, whereas a dark blue was at first the favourite foundation at Rome and Ravenna. While the most splendid coloured marbles clothe the walls of S. Sophia, mosaics on gold ground cover all its vaults and domes, as they cover St. Mark's at Venice; and Salzenberg, who saw them freed from the whitewash under which the Turks have partially buried them, is eloquent over their rich and varied effect of flashing light and deep but glowing shadow.

This use of gold has been held to be of Oriental origin, and we are introduced here to the question of the debt of Byzantine art to the East. The evidence for this appears most strongly in Byzantine love of costly material, of glancing lights, and of splendid colours. It was probably through her gold, her jewels, and especially her brilliant figured and embroidered stuffs, that the East made her influence felt. Nothing was a greater proof of the decline of pure classical taste than the elaborately adorned robes in which Byzantine courtiers enveloped their persons. These robes corresponded with the minute court ceremonial and multitudinous etiquette, which to a modern eye would have seemed the most Oriental thing at Byzantium. The cumbrous shape of these garments, and the weight and stiffness of the jewels and gold with which they were loaded, rendered the true classical treatment in art of the human form and drapery no longer possible, and the taste which delighted in these things demanded a similar splendour and varied glance in the decoration of walls or of the pages of manuscripts. There is a certain Oriental sound in the descriptions given by Byzantine writers of the glories of their churches, and of the apartments in the imperial palaces. We seem to be introduced by them into some magician's cave of an Eastern tale, where all the wealth of the underworld is heaped together in formless profusion. Such accounts might easily convey a very false idea of the *style* of Byzantine decoration which, in all its richness, seems never to have lost itself in the romantic exuberance of Oriental forms. There is, indeed, comparatively little of Eastern extravagance and Eastern waywardness of ornamentation in good Byzantine work. Colours are brilliant, and gold employed to excess, but classical tradition still rules the forms, the severity of which is a continual surprise to students of Byzantine manuscripts, mosaics, and gold-enframed enamels.

There are, as we have said, traces of Eastern influence in the decorative carving of the Byzantines. Their acanthus, though designed with the most delicate taste, exhibits a poverty in relief, a thinness in the forms, and a sharp angularity in cutting, which is opposed to classical feeling. Greek foliage is largely modelled, and full and rounded in profile; it is the work of born sculptors. This fine plastic quality is absent from Byzantine carving, which tends to become mere flat decoration. In this it is possible to see the influence of the elegant but *feminine* Oriental ornament; but it may be sufficient to ascribe it to the comparative absence of sculpturesque feeling in the Byzantines. It is noticeable how little the effect of their architecture depends on prominent mouldings, and how much upon colour and carving which, though actually in relief, is little more than design. In this characteristic their architecture resembles that of the Arabs, which was so closely connected with it, and it differs from that of classical antiquity and of the Western nations. The first adorn their architecture with graphic, the latter with plastic decoration. While the Byzantine and Moorish builders cover their surfaces with gold or rich incrustations, or with an exquisite maze of interlacing ornament, the Greeks and mediævals relieve theirs by free standing statues or reliefs, by bold cornices and bosses, and by an interchange throughout of projection and recess.

In their treatment of other foliage, the Byzantine carvers preserve the same characteristic sharpness, and delight in spreading a network of flatly-treated leaves and tendrils over a capital or a panel. The introduction amidst this foliage of animal forms is a further feature which may be connected with the East. An excellent example from the sixth century occurs in some of the panels of the ivory chair of Bishop Maximian at Ravenna, the Byzantine character of which we may assume without committing ourselves to the opinion that Ravenna decoration generally is Greek. These panels are covered with an arabesque of vine leaves, in the midst of which are playing all sorts of birds and four-footed creatures, charmingly natural in form and action, but drawn without any observation of scale, so that the raven stands as tall as the ox, and the squirrel's tail is as formidable as the lion's mane. In designs like these, which were specially favoured when the iconoclastic movement had excluded the sculptor from the treatment of the human form, Byzantine art seems to occupy a middle place between classical and Oriental. It is freer than the first, but preserves a severer sense of form than the latter. The decorative forms of the

Byzantines are imitated from nature, and are something more than mere graceful but aimless lines and flourishes like those of the Persians and the Arabs. Their animals, again, are true to life, and not grotesque and fantastic like those in Eastern stuffs, or, we may add, in the decoration of the Celtic and Teutonic peoples of the West. Whether or not there is a connection, either belonging to the prehistoric time, or instituted later through commerce, between these creations of Oriental and of Western fancy, is a question which cannot be touched on here. Professor Unger has gone so far as to suggest that the leaf and animal ornament of the Irish manuscripts may have been an importation from Byzantium.[1] It is true that there occasionally occur in these illuminations motives, such as a peacock drinking at a vase, which are certainly Byzantine, and testify to the early commercial connection between Constantinople and the Western isles. But the whole style of these Celtic manuscripts, as well as the style of northern ornament generally, culminating in the carved Romanesque capitals, is at the opposite pole to that of the Byzantines. The former is weakest when imitating nature, and strongest when the artist follows the dictates of a quaint but exuberant fancy. The Byzantines, heirs of antiquity, never abandon the results of their classical training, always preserve a grasp of form, and know what it is that they are doing; so that their art in its decline, though it stiffens into rigidity, does not lose itself in vagueness.

Glancing at Byzantine art as a whole, we are struck by two special characteristics. The first is its monotony and want of freshness. In its decline it passes into utter lifelessness and rigidity, and even through the period of its bloom, though it is free from this reproach, it still shows the want of that independent study of nature and openness to new impressions, which made the life of the art of Greece and of

[1] Unger, in Ersch und Gruber's *Encyklopädie*, § 1, Th. 85, p. 19.

mediæval Italy. The second is its extreme technical finish; if technical perfection in decorative art can make up for the want in it of true vitality, Byzantine art is beyond reproach. Nothing more exquisite in delicacy and taste than a piece of Byzantine *cloisonné* enamel, or a small elaborately finished figure from a manuscript, or an ivory carving of a good period, can possibly be imagined, and it was through its fine technique that the art was able to survive so long, and to exercise so potent an influence on the more vigorous, but less cultured, art of the West.

It would be unjust, however, to deny to Byzantium the credit of having produced artistic creations of high and independent value. We have already noticed, in connection with the basilica, that phase of Christian art which may be suitably called *epic*, wherein the majesty of the Church finds expression in forms that have borrowed the dignity of the antique. This aspect of Christianity was especially prominent at Byzantium, where religion, always in closest connection with the court, wore the imposing garb of an imperial institution. We are fortunate in possessing in the British Museum a specimen of Byzantine design which realises the ideal of this majestic style. It is an ivory relief of an angel on the leaf of a diptych, and for noble dignity and grace could hardly be matched. The figure stands full face, and is finely posed at the top of a flight of steps. The forms are full and roundly treated, the drapery simple and classical, the raised hand, which holds a sceptre, well drawn and full of action. Equally fine are some mosaics of uncertain date, which still remain more or less visible in S. Sophia. One represents a colossal archangel as noble as that in the ivory relief just mentioned, the other, which is in the vestibule, a seated figure of Christ of benign but majestic appearance. At His feet is prostrating himself an emperor, whom some recognise as Justinian, while the gold surface behind the throne is broken by two medallions containing heads of Mary and the Archangel Michael. If the reproductions of Salzenberg give a true idea of these works, their dignity and charm might win for them a place beside the finest designs of Greek and of Renaissance artists. What must have been the aspect of the interior of S. Sophia when shapes like these stood forth from the gold ground of the mosaics, and when the eye, wearied with following the intricate foliage carving of the capitals, or sated with the varied richness of the marbles, could rise toward the solemn cupola spreading there its peace over all, and meet the monumental forms of angel or of saint, and the Christ enthroned aloft on the rainbow, the Judge of all the earth!

The problem of the Christian domed church, so far at least as its interior treatment is concerned, receives in S. Sophia its full solution. This problem was, as we saw, to unite a hemispherical dome like that of the Pantheon with a ground-plan which should offer the convenience of the basilica. This union was secured in S. Sophia, where the simple but grand effect of the all-embracing dome is preserved, while the convenience of the plan and the internal arrangement of the spaces are in no way sacrificed. In later Byzantine churches this special effect is no longer aimed at. That correspondence between the ground-plan of the building and the form of the dome which the architect of the Pantheon secured with such thoroughly Roman directness, is preserved in S. Costanza, S. Lorenzo, and S. Vitale, and is retained unbroken, through the introduction of the half domes, in the more complicated arrangements of S. Sophia. From this time forward, however, all such correspondence is lost, and cupolas are used upon buildings, the plan of which has no more real connection with domical than with any other form of roofing. Circular and polygonal ground-plans for large churches now pass out of use,[1] and the dome is used in connection with square and even cruciform plans where its function is far less pronounced than in the earlier structures with which we have been dealing. Domes are used henceforth either for simple covering, or to provide spaces for decoration, or to emphasise some important part of a building, or again—and this is the noblest use of the dome in modern times—for exterior effect. They remain convenient and beautiful but not specially significant, until in the hands of Brunelleschi, Buonarotti, and Wren they receive a new character as imposing external features.

This change in the use of the dome begins even under Justinian. Constantine had erected in his new city a church dedicated to the apostles, and destined for the mausoleum of the imperial family. It was of the form of a Greek cross, and was perhaps the first important Christian building presenting the cruciform ground-plan, though its destruction leaves the well-known mausoleum of Galla Placidia at Ravenna the earliest building of the kind which still remains. Over the crossing there was probably a cupola, which had there the suitable function of emphasising that portion of the church where the imperial sarcophagi were deposited. Justinian, who rebuilt this edifice on the old ground-plan, added to the central dome four others, one over each arm of the cross, thus giving the first example of the plan familiar to all through St. Mark's at Venice. It is common again, in later Byzantine churches, to find five domes arranged over a square ground-plan like the five dots on a die—over the centre and the four corners—while a row of cupolas is sometimes used to cover a vestibule, or even the nave of a church. In all these cases, though the beauty of the domes taken singly remains, yet the multiplication of them is entirely destructive of that which is finest in the cupolas of the

---

[1] Polygonal plans remain in use for small buildings like Baptisteries.

Pantheon, of S. Vitale and of S. Sophia,—the unifying effect of the dome as the all-embracing feature of the building.

The cupola, as used to accent an important portion of an edifice, finds a suitable place over the crossing of the cruciform churches of the middle ages. The fact that a tower with a cross-vault underneath, and later on a *flêche*, or slender spire, appears just as frequently in this position, shows that there was no feeling that the dome had any special æsthetic value.

It is only in its exterior aspects that domed construction made real advances after the time of Justinian. The domes of S. Sophia and of the Pantheon are in their lower portions heavily charged, and while the external aspect of the former is insignificant, the latter is, for half its height, concealed from view altogether. It is not till the dome is elevated on a free standing drum above the building that it makes any impression from a distance. This feature, which is the secret of the imposing effect of Sir Christopher Wren's masterpiece, and to a lesser extent of the dome of St. Peter's, is first seen in the 'Dome of the Rock' at Jerusalem, and appears in many Byzantine churches. It receives, however, a further development at the hands of the Renaissance builders. That the dome which, with the Romans, was purely an *internal* feature, assumes such great *external* importance in Renaissance buildings, is an interesting fact in architectural history.

The suitability of the dome as a feature in Christian architecture claims a concluding word. We have seen that it has a very distinct æsthetic significance, and is through this peculiarly expressive of the Roman genius, though it is perfectly true that a certain fixed and contemplative character, which belongs to it, suggests naturally a connection with Oriental ideas. This was all in its favour as the dominant feature in the architecture of despotic and semi-oriental

Byzantium, where spiritual freedom had so small a field of exercise. It explains, too, its reception among the proud votaries of the culture of the Italian Renaissance. But this character is opposed entirely to the genius of Christianity, with its perpetual demand, its limitless promise. The dome directs toward the centre, it bars egress. To the religious architecture of the West was given the mission to create forms which should direct the beholder forth from himself and towards the infinite, the spirit of which should be aspiration; and Christianity does not speak through the domes of Byzantium, Rome, or London, but through the tower of Strasburg and the vault of Beauvais.

CHAPTER V.

CHRISTIAN ARCHITECTURE IN THE WEST, FROM THE EARLY CHRISTIAN TO THE MEDIÆVAL PERIOD.

IT would be impossible within the limits of a single chapter to do justice to the rich and varied history of ecclesiastical architecture through the numerous schools of mediæval Europe. For the more strictly architectural facts belonging to this period the reader must be referred to the many excellent handbooks now so accessible; and what will rather be attempted in this place will be to penetrate, so to say, beneath these strictly architectural facts, and to discover if possible the underlying causes which at each epoch have determined their character. For religious, national, and social ideas and movements have a most important part to play in determining the aspect and character of buildings, and that is but a limited interest in architectural history which attaches itself only to constructive forms and the affiliation of successive styles of decoration. The buildings of old time arose in response to living needs and aspirations, and as the body to the spirit so their structure and ornamentation answered to the life which went on within them. If therefore we are to represent to ourselves as a vital thing this wondrous growth and unfolding of Christian architecture, it must be by realising something at any rate of the spirit of the communities for whose service it came into being. It is apparent at a glance that the early Christian basilica is in architectural character widely different from the Romanesque minster, while the Gothic cathedral has also its distinct and independent position. These extraordinary differences, appearing while the main elements of the general plan remain the same throughout, can only partially be explained on mechanical principles. It is true that we may trace a *constructive* development of the forms of the basilica from the fourth to the fourteenth centuries, and in this resides the purely architectural interest of the period. It is none the less true, on the other hand, that it was the tone and spirit of the Church and of society generally which were throughout the dominating factors. How these operated in the early Gothic period will be noticed on a subsequent page; our present concern is with the architectural history of the earlier mediæval period in its connection on the one hand with the inner life of the Church, and on the other with the momentous outward events brought about by the Teutonic invasions of the West.

In the early patristic writing known as the *Apostolical Constitutions* there is an interesting picture of the interior of a Christian church at the time of solemn service, and the general tone of the passage has much significance for the subject of the present chapter. In this curious description a fanciful analogy is dwelt upon between the church with its presiding officials and its congregation, and a great ship equipped with officers, with a disciplined crew, and with a body of passengers who have to be kept in strict order. The bishop is to call the assembly together as the captain of the vessel, charging the deacons as mariners to prepare places for the brethren as for passengers, with all due care and decency. The building and the gathering are then described as follows:—

'Let the house of assembly be long in shape and turned towards the east, with its vestries on each side at the eastern (entrance) end, after the manner of a ship. Let the throne of the bishop be placed in the midst, and on each side of him let the presbytery sit down, while the deacons stand beside with closely girt garments, for they are like the sailors and managers of the ship. In accordance with their arrangement let the laity sit on the other side with all quietness and good order, and let the women too be in a place apart and sit in order keeping silence. . . . Let the porters stand at the entrances of the men and give heed to them, while the deacons stand at those of the women, like shipmen . . . and if any one is found sitting in a wrong place let him be rebuked by the deacon as manager of the foreship and removed into the place proper for him, for the church is not only like a ship but also like a sheepfold, and as the shepherds place all the brute creatures distinctly . . . so is it to be in the assembly. Let the young men sit by themselves if there be a place for them, but if not let them stand upright, but let those already advanced in years sit in order, and let the children stand beside their mothers and fathers. Let the younger women also sit apart if there be a place for them, and if not let them stand behind the elder women. Let those women who are married and have children be placed by themselves, while the virgins and the widows and the elder women stand or sit before all the rest, and let the deacon be the disposer of the places that every one that comes in may go to his proper place, and not sit at the entrance. . . . In like manner let the deacon oversee the people, that nobody may whisper nor slumber nor laugh nor nod, for all ought in the church to stand wisely and soberly and attentively, having their attention fixed upon the word of the Lord. After this let all rise up with one consent, and looking towards the East, after the catechumens and penitents are gone out, pray to God eastward. . . . As to the deacons, after the prayer is over, let some of them attend upon the oblation of the eucharist, ministering to the Lord's body with fear. Let others of them watch the multitude and keep them silent . . . (During the celebration) Let the door be watched, lest any unbeliever, or one not yet initiated, come in.'[1]

The above passage, which may be referred to a date in the third century, stands almost alone among early patristic writings for its distinct presentation of the Christian meeting-house and its arrangements. For our present purpose the point which invites attention is the almost painful effort

[1] *Apostolical Constitutions*, ii. 57. (Translated in the Ante-Nicene Library, vol. xvii. p. 83 ff.)

made to secure *order* on the part of the congregation; in the interests of order the whole assembly is under the absolute control of the officials, and every section has an appointed place! The way is prepared for the outward development of church organisation in the middle ages, and for the unyielding assertion of the hierarchical principle, both in the relation of individuals to ecclesiastical superiors, and in that of subordinate communities and churches to the central authority at Rome. With the outcome of that special part of church organisation which has relation to ritual and to ceremonial observances connected with places and seasons, we have in this place comparatively little to do, for this influenced rather the interior fittings than the general form of the churches. But with the effort of the Church after ORDER the case is different. Here we have to deal with a deep underlying principle that was of vital importance in the development of architectural forms, and it belongs to the argument in this chapter to emphasise the view that this principle was in great part an inheritance from Rome. It is a remarkable illustration of the support received by the Church in her work of ecclesiastical organisation by this inheritance from Rome, to find that the great conventual establishments of the middle ages, in which her power chiefly resided (and which were the homes of architecture and the arts), were both in their origin and their relation to each other and to the Papal See, not a little like Roman colonies. These we know were 'outposts of the Empire,'—wedges thrust into a hostile region to be the means of reducing and holding it beneath the Roman sway. In like manner the monasteries were Christian colonies accomplishing the same work in the service of the Church. If all colonies were, as Cicero says, 'small images and models of the Roman state,' so all monasteries conformed to a certain general pattern. As the Roman city had been the central pivot of a vast system of provincial organisation, so now the Roman Church, inheriting the imperial tradition, made herself the religious metropolis of the West, linking the scattered ecclesiastical establishments into one grand unity, in the administration of which Roman maxims of government still lived on. Throughout this ecclesiastical system reigned the Roman spirit of order, *and it is the working of this same spirit in the sphere of architecture which makes the difference between the Romanesque minster and the early Christian basilica.* The truth of this will be seen when we glance at the main points of contrast between these buildings, but before proceeding to this it will be necessary to introduce a brief digression.

In insisting upon the Roman character of early mediæval architecture it must not be forgotten that this architecture belonged in great part to Teutonic, or at any rate largely Teutonised lands—Saxony, the Rhineland, Northern France, Lombardy,—and no general argument on the subject before us can be complete without a previous consideration of the question, What effect upon Christian architecture was produced by the Teutonic invasions of the West?

The circumstances under which the development of Western buildings proceeded were widely different from those which surrounded the Byzantine architecture noticed in the last chapter. While the architects of the Eastern Empire, secure behind the impregnable bulwarks of Constantinople, could work out to their utmost logical consequence the forms with which they started, their brethren of the West were subjected to another set of conditions altogether which might be expected to exercise over their work a most powerful influence. The main facts in the history of the Western provinces of the Roman Empire from the fifth century onwards are familiar to all. Not fifty years had elapsed since the death of Constantine, before the

Goths had crushed the legions of the Emperor Valens at Hadrianople by a blow heavier than any Roman army had felt since Cannæ, and the inroads of the barbarians into the Empire had begun in earnest. During the next hundred years the West was swept by wave after wave of Teutonic invasion, till by the beginning of the sixth century the Roman Empire of the West had to outward appearance passed away, and the ruling powers in Illyricum and Gaul, in Italy and Spain, and in the province of Africa were no longer the Imperial præfects, but kings of the Goths the Franks and the Vandals, while a little later a body of wilder Teutonic invaders had possessed themselves of the once important Roman province of Britain. In the course of the seventh century the Saracens overran Syria and destroyed there the flourishing Christian communities, the remains of whose architecture are of such unique interest to the modern student. Advancing westward along the Mediterranean coast, the tide of Mohammedan invasion overwhelmed the African province,—one of the most busy and populous seats of Christianity,—and, overflowing into Spain, drove the Christian population for a time into the remote corners of the peninsula.

What, we may ask again, was the effect of these vast historical changes upon the future of Christian architecture?

The Teutonic invasions of the West assume different aspects, according as we dwell upon the things which were destroyed, or upon those which were left intact. The invaders were barbarians, and could show themselves on occasion as lawless, insolent and cruel as any of their kind, but their barbarism was greatly tempered by the fact that they were in many cases no strangers to Roman culture and religion, and had grown up with an instinct of reverence for Rome and an aptitude for the reception of whatever Rome could teach them. The invaders shattered the armies of the Empire and sacked her cities. The soldiers of Alaric 'filled the streets of Rome with dead bodies,' stripped her palaces and sold her citizens into slavery. Teutonic chieftains wrested from the Emperor his power of command, and were ready to defy to his face, in moments of irritation, the impotent wearer of the purple. But against all that, in the deeper sense, was Rome—they were powerless. All these outward things that could be touched and handled they might destroy or seize, but before the ideas, the culture, the art, the institutions, and above all the religion of Rome they were irresistibly compelled to bow. The often-quoted words of one of the ablest barbarian leaders bring out into the clearest light their attitude towards Roman civilisation. 'When I was young and eager in mind and body,' said Athaulf the successor of the great Alaric, 'I had at first vehemently desired to blot out the Roman name, and make and call all that was Roman the kingdom of the Goths alone, till Romania should have become Gothia, and where there had once been Cæsar Augustus there should now be Athaulf. But taught by long experience of the unbridled savagery of the Goths and fearful of depriving the State of those laws without which it would cease to be a State at all, I chose rather to make it my glory to restore anew and to exalt the Roman name through the vigour and strength of the Goths, so that I should be known to posterity as the author of the restitution of Rome since fate had not given it to me to be her remover.'[1]

Among the institutions of the Empire which loomed so majestically before the eyes of the strangers from the North, none was more imposing than the Christian Church. The influence which the Church possessed in virtue of her association with Rome was rendered irresistible through a certain natural affinity in the Teutonic nature for Christian

[1] Orosius, *Hist.*, vii. 43.

ideas, which explains the rapid and complete conversion of the invaders even before they had violated the boundaries of the Empire. Striking is the scene described by Orosius at the sack of Rome by Alaric. A Gothic chieftain had invaded the house of an ancient virgin who had the care of a church. On his demand she produced a treasure of massive objects of gold and silver plate, but warned him that they were the sacred vessels of St. Peter. The barbarian in awe sent to inform Alaric, who ordered the spoil with its custodian to be conveyed in safety to the church of the apostle. Borne upon the heads of the gigantic barbarians the glittering vessels were transported through the streets of Rome, while a guard with drawn swords encircled the procession, and Romans and Goths sang sacred hymns in concert as it passed.[1]

To the Church Christianity was no question of race though it was one of orthodoxy, and Greek and Jew, barbarian and Scythian were all alike before her. The continuity of religion was of greater moment than the disruption of political relations, and Orosius who was a Roman citizen of Spain protests that 'if the barbarians had been sent into the Roman territory only for this one reason that it should be possible to see on every side, throughout the East and West alike, the churches of Christ filled with an innumerable and varied multitude of believers,—Huns and Suevi, Vandals and Burgundians,—it would be right to praise and magnify the mercy of God in that through the ruin of the provincials he had brought so many nations to a knowledge of the truth to which otherwise they would never have attained.'[2]

This attitude of the barbarians towards Roman civilisation and towards Christianity, is an important fact in connection with the history of Christian architecture in the

---

[1] Orosius, *Hist.*, c. 39. [2] *Ibid.* c. 41.

Teutonic kingdoms of the West. The one exception to this attitude was in the case of the English, who had never been touched in their native seats by Roman influences, and whose conquest of Britain involved a destruction of the inhabitants which was without a parallel in other lands. While therefore the native Christians were driven into the hilly regions where their representatives still remain, Britain generally relapsed for a time into paganism, till Christianity was imported into it afresh from Rome at the end of the sixth century. The complete abolition of Christian institutions in the lands conquered by the Saracens is also of course to be taken for granted, but except in these two instances the barbarian conquests did nothing to retard the mission of the Church. The continuity of religious work under Roman and barbarian rule is well illustrated by a passage in Gregory of Tours (died 594), in which at the end of his history he enumerates the churches erected by his predecessors in the bishopric of Tours from the third to the end of the sixth century. The account of the early times is additionally interesting as showing how similar the conditions of church-building were in the provinces to those which we have seen prevailing at Rome.[1]

Gatian the first bishop was sent from Rome in the time of the Emperor Decius. . . . He had a few followers who met in crypts and caves of the earth because of the persecution. . . . He was buried in the cemetery which belonged to the Christians.

The second was Litorius a native of Touraine; he erected the first church at Tours, for the Christians were now numerous and the first basilica was formed by him out of the dwelling of a certain senator.

The third was St. Martin, consecrated 374 A.D.; . . . he built a basilica in the monastery afterwards called the

[1] Greg. Tur., *Hist. Franc.*, x. 31 *seq.*

Greater, and visiting the villages round about, destroyed the shrines, baptized the folk, and erected numerous churches. His successor Briccius followed his example and erected moreover a small basilica over the body of St. Martin, in which he was himself afterwards buried.[1]

The fifth bishop also built churches in the districts round about, while the sixth, Perpetuus, a very rich man and of senatorial family, pulled down the small basilica of Briccius and erected a larger one, very beautifully built, into the apse of which he translated the body of St. Martin. A baptistery is mentioned as existing in his time. He built also several churches in town and country, including the basilica of St. Peter. On his death he left his possessions by will to the churches.

The seventh bishop had similar rank and wealth which he expended on church-building. *In his time Clovis already bore rule in some of the cities of Gaul* (c. 490), and the prelate suspected by the Goths (who had preceded the Franks in the conquest of south-western Gaul) of an intention to submit to the rule of the Franks, was condemned to exile at Toulouse where he died. His successor shared the same fate. In the time of the ninth bishop, Clovis the king returned victorious to Tours after the defeat of the Goths. After him co-bishops, Theodorus and Proculus, were appointed at the command of the blessed Queen Clotilda, whom they had followed from their native Burgundy.

Leo, the thirteenth bishop, was an artificer in wood and made towers covered with gilding, of which remains existed in Gregory's days. He was skilled too in other works. . . .

Baldwin became the sixteenth bishop, having previously

---

[1] This was the origin of the famous church of St. Martin at Tours, one of the most important in Christendom.

been chancellor to King Clotaire. The seventeenth, Gunthar (we notice how Teutonic names occur), was often employed on political missions by the kings of the Franks. In the time of Eufronius, Tours with all its churches is consumed by a great conflagration. Two of these he rebuilds, and restores and covers with lead (stanno), *at the expense of King Clothair*, the basilica of St. Martin (*i.e.* the famous church). In his time was erected the basilica of St. Vincent as well as district churches.

The nineteenth bishop was Gregory himself (c. 573-594). The list of his works comprises the rebuilding in an enlarged form and with more lofty roof of the cathedral of Tours, the restoration and decoration of the basilica of St. Perpetuus, and the erection by it of a baptistery, together with the dedication of many churches and oratories in and about Tours, which he will not weary his readers by describing.

This account is sufficient to show how little the settlement of the barbarians in the Empire interfered with the church building which was one of the primal cares of these wealthy bishops. It is said by Fergusson that about the sixth or seventh centuries there 'occurs an hiatus in the architectural history of Western Europe, owing to the troubles which arose on the dissolution of the Roman Empire and the irruption of the Barbarian hordes.'[1] This is only true as regards the *sequence of existing monuments*, for these, owing to the constant rebuilding which went on in the early middle age, are in almost every case of considerably later date. Literary records however, like the passage just quoted, make it clear that the barbarians had no intention of interfering with architectural operations, but continued the patronage of the Church formerly exercised by the Roman government. It is true that these invasions and conflicts of tribe

[1] *History of Architecture*, i. p. 396.

with tribe threw the country into considerable disorder, and it is true too that a barbarian chieftain might not seldom, through accident or through irritation, burn a church or harry a monastery, but the bishops and abbots were always masters of the situation and knew how to make the spoiler restore fourfold what he had destroyed. There is no question therefore that church building went on steadily through the centuries between the sixth and eleventh, under the same conditions which surrounded it in the fifth. There is an hiatus in the monuments, but although the age of the basilicas is separated from that of the early Romanesque churches by a long period to which but few existing monuments can be ascribed, it would be a mistake to assume that they represent entirely different developments of architecture. The basilica belongs to the Roman world, the Romanesque church arose in lands occupied by Teutonic invaders; but it does not follow that the basilican style died out with the Roman authority, while the Romanesque or the mediæval style represents a new beginning on Teutonic lines in independence of the old. Had we a charm to conjure back into existence for a moment all the ancient churches which have succeeded each other from generation to generation in some famous city like Tours, could we see Briccius' basilica of St. Martin over-against the five others which were successively erected on the same site at various dates between the fifth and the thirteenth centuries, could we reconstruct all these edifices which fire or the crowbar or decay have crumbled to ruins, and set them side by side in order, we should have before us no signs of an hiatus or a sudden change, but the successive steps in a long course of unbroken development which we can trace, though by no means so clearly as we could wish, in ecclesiastical records.

Certain monuments which we do possess from this period illustrate in a striking manner this continuity of architectural traditions from the early Christian to the mediæval epoch. In the midst of various uncertain or fragmentary structures, two important buildings stand out as representative of the architecture of the two greatest of the Teutonic kingdoms founded on the ruins of the Western Empire. Both are sepulchral chapels of great barbarian monarchs, and as such might be expected to show stronger traces of a traditional northern style than any other monuments. One is the tomb of Theodoric the Ostrogoth at Ravenna, the other the church erected by Charles the Great at Aachen, partly for his palace chapel and partly for his sepulchre. The latter has already been referred to as an example of domed construction on Roman principles, and there is little in either building which is not in accordance with classical traditions. The Mausoleum of Theodoric is imitated from Roman mausoleums like that of Hadrian, but exhibits ornament of barbarian type, not unlike that which we find in the early Norman churches, while the Minster at Aachen is in structure detail and ornament throughout a Roman building. With these edifices before us we need hardly be told of Theodoric's admiring care for classical remains, or the zeal with which Charles sent for workmen from 'all the regions on this side the sea'—*i.e.* from Gaul and Italy, and imported costly marbles and other material from Ravenna and from Rome.

It is possible now to fix the true connection of the northern tribes with the development of Christian architecture in the West. We have seen that they were ready from the first to carry on in the old course the traditions of early Christian building, and that no idea of sweeping away the old and making a new departure ever dwelt for long in their minds. Yet we have seen on the other hand that both the Romanesque church of the eleventh century and the Gothic cathedral of the thirteenth, are buildings in many ways strikingly unlike the basilica, and though retaining the same general plan have assumed forms of independent constructive and æsthetic interest.

What part, it may naturally be asked, did the Teutonic genius play in this development? The following sentences contain in brief the answer to this question.

I. The barbarian tribes possessed no native style of architecture and no solid technique, nor do they appear to have brought with them into the Empire any distinct architectural tendencies, but were content to learn and use the style and technique of the Roman population of the countries which they overran.

II. The constructive forms which were added to the basilica to produce the Romanesque church of the early middle ages were in no case of barbarian but of Roman origin, and there is nothing distinctively northern or Teutonic in the way in which they were at first dealt with by mediæval builders.

III. While the attitude of the barbarians towards Roman architecture was, so far as the form and the structure of the buildings are concerned, purely receptive, they employed from the first in the regions which they made their own, certain peculiarities in building for which there is no classical authority, and, more especially, a system of ornamentation strongly tinctured with northern feeling, so that, speaking broadly, the early mediæval church is in form Roman but in decoration Teutonic.

IV. In the Gothic architecture of the later mediæval period the Teutonic spirit for the first time distinctly asserts itself, and all that is most characteristic in that great development of architecture is the expression of the romantic spirit of the North.

A few words in illustration of these statements will be sufficient for the present purpose.

In the first place there is no evidence to show that
either the tribes who inhabited the West before the Romans
subdued it, or those who pressed into it from the North and
East in the fourth and following centuries, knew of any
style of building except simple constructions of wattle and
clay or of wood.

Of the Gauls Strabo writes that 'they dwell in houses of
considerable size, constructed of planks and wicker work,
and covered with heavy thatched roofs of conical form,'[1]
and he records of the Britons that 'forests are their cities,
for having enclosed an ample space with felled trees they
make themselves huts therein and lodge their cattle.'[2] In
the case of the Teutons there is linguistic evidence of the
same practice in the prevalence in the old Teutonic tongues
of forms of the one verb 'to timber' for all kinds of building, while the accounts which Tacitus and later writers give
of the nations beyond the Rhine describe them as living in
scattered communities with no regularly constructed buildings. 'The houses of the Germans,' writes Herodian in the
third century, 'are very soon burned down because they are
mostly of wood.'[3] We possess a curious account of an
embassy sent from Constantinople in the year 448 to Attila,
king of the Huns, at a time when he was sojourning in the
land of the Goths beyond the Danube, and the description
of his residence gives an idea of the style of building prevalent in those regions.[4] 'It was put together of well-
smoothed beams and planks and surrounded by a palisade
of wood, not intended for defence but for adornment.' This
building was also furnished with towers while the lodging of
Attila's consort was more elaborate in construction and was
ornamented with carving; the whole, to quote Schnaase's
words, presenting the character of 'a very fully developed

[1] iv. 4, § 3.
[2] iv. 5, § 2.
[3] Herodian, vii. 2, § 3.
[4] Script. Hist. Byz., pars vi. p. 187.

timber architecture with a probably fantastic but tolerably rich decoration.'[1]

We know from records as well as from existing remains that the architecture of the Saxons and other Teutonic invaders of Britain was also confined to wooden constructions, while in Norway there are timber churches dating from about the twelfth century which represent the ancient style of the Scandinavians, while their curious ornamentation, consisting of dragons and interlacing bands, may give us an idea of that seen in the fifth century on the dwellings of Attila.

This native style did not die out upon the introduction of Roman civilisation, for we know that in Gaul and in Britain there was kept up until the middle ages a primitive manner of construction in wood or wattles called 'Gallic' or 'Scottish' which was contrasted with the substantial building of the Romans. Thus Bede speaks of a church erected in the Celtic style and not of stone (*more Scotorum, non de lapide*), and in the seventh century we hear of a basilica erected 'not after the Gallic manner of pieces of wood and beams,' *minutis ac rudibus,* but of squared stones, while at the same period Abbot Benedict of Wearmouth sends to Gaul for stone-masons to erect for him a 'stone church after the manner of the Romans.'[2] The architectural efforts of the non-Romanised inhabitants of the West were confined therefore to wooden constructions, slight and easily destructible, though often most ingeniously put together and elaborately decorated with carving and with colour. For solid structures of brick and stone they had recourse to classical models and to the 'Roman manner.'

This 'Roman manner' carries us back to the days of Constantine and his successors when, as we have seen, efforts were made to arrest the decay of ancient practice in the arts

---

[1] *Geschichte der bild. Künste,* 2 Aufl., iii. 509.
[2] See the references in Schnaase, iii. 521 ff.

by the establishment of provincial schools of architecture and decoration. In these schools the classical tradition was handed down unbroken to the middle ages, and it was to these that the barbarian chieftains turned whenever they desired to raise any durable monuments. The towns of northern Italy and of Gaul naturally contained a large number of these classically trained artificers whose services were invoked by the Goths, the Franks, and the Lombards. It was by the hands of these that the bishops of Tours built their churches under the Frankish kings, and they were sent for, as we have seen, from Gaul to Britain, and also summoned from Italy into the regions of Germany. After a time the men of the new races learned the same craft and boasted that they could build in the Roman style without the aid of the conquered population, as in a Gothic inscription from the South of France, testifying that 'this work, which no stranger of Roman race came to fashion, has been accomplished by a man of the barbarian stock.'

It follows from what has now been said, that the Teutonic invasions of the West did not exercise any definite determining influence upon Christian architecture in the early mediæval period. The church of that period was not a creation of the Teutonic genius, and those special features in which it differs so markedly from the earlier basilica and schola are to be explained, not as an importation from the North, but as the result of a normal development upon lines already indicated. The Romanesque church, stately but often heavy, and always severely logical, is a testimony to the power of the grand principle of Order inherited and maintained by the Church as the outcome of the Roman system of government and law. Hence the significance of the name 'Romanesque,' which some have been tempted to drop for the vaguer term 'early mediæval,' or the meaningless 'round-arched Gothic.' To us 'Gothic' is in itself a perfectly meaningless word, but, by a convention of language, it is used to express a certain very distinct phase of Christian architecture, and in this connection everybody understands it. 'Round-arched Gothic' is enigmatical, for the round arch is a characteristic Roman form, and yet no one would speak of 'Roman Gothic.' Those who employ this expression probably are of opinion that 'Gothic' has a real ethnological meaning as equivalent to 'northern' or 'Teutonic,' and that the round-arched style of the early mediæval period is, equally with the pointed style of the thirteenth century, a northern production. It is the thesis of this chapter, however, that this round-arched style is, in its essential character, a continuation of Roman traditions, and for the support of this, appeal may be made to the Syrian churches described by de Vogüé, which exhibit an advance from the basilica in the direction of Romanesque forms strikingly parallel to that exhibited in the West, though upon soil which no early Teutonic invader ever visited. In the next few pages it is proposed to notice some of the salient points in the constructive development of the early mediæval church, and to fix, as far as possible, the first appearance of characteristic Romanesque features.

In preceding chapters, the various 'root-fibres' of Christian architecture have been traced back to their probable origin in buildings familiar to the Greco-Roman world. We have seen reason to connect the church, or certain features of it, with private halls, lodge-rooms, memorial cellæ, or even underground crypts of the cemeteries. What is required, however, is to discover, if possible, some one guiding principle of the earliest Christian architecture, some general scheme that would always be adhered to. As a fact, a guiding principle of this kind did really exist, and is not far to seek. Undoubtedly, the true germ of the Christian church was *an oblong interior terminated by an apse.* It appears to have been a well-understood tradition among the early Christian builders, that whatever might be the special features of their design, this fundamental idea must be followed, and this tradition is the determining influence in the architecture of the age of Constantine. Those who make much of the pagan basilica as the true form-giving element at this important epoch, must deny or pass lightly over this previous tradition, and a criticism of their views is reserved for the Appendix to this volume. It is sufficient here to insist once more upon the fact that the Christian basilica of the fourth century could not have derived its apse, the essential feature of its plan, from the pagan basilica, for this building, which '*non é di sua natura absidata,*' had no such feature to lend. The apsidal termination, with the oblong interior in such exact accordance therewith, must have been previously established, and no more simple and natural origin can be suggested for it than the city *schola* of religious brotherhoods, after the pattern of the Pompeiian buildings shown in plan on page 52. It is on the hypothesis that this form had become established by the end of the third century as the most suitable for the Christian church, that we can best explain the characteristic features of the architecture of the period which succeeded. Especially welcome is the explanation offered by this hypothesis of one feature which has always been somewhat of a stumbling-block. This is the appearance, in the Christian basilica, of a plain wall above the colonnades of the nave, where in the pagan basilica there is placed a gallery. This plain wall has been severely criticised as an architectural crudity, and even quoted as the first sign of the barbarism which was to destroy the purity of antique forms. Unquestionably the plain wall, as we see it, for example, in St. Paul's at Rome, or as it must have appeared in Old St. Peter's, is of overpowering extent when compared with the colonnade on which it rests, and its monotonous surface is destitute of all architectural breaks and subdivisions other than the simple mouldings which frame the mosaic pictures. A reason for its appearance in the Christian church may be found in the desire to secure extensive spaces for the display of these sacred pictures, the use of which has been illustrated from Paulinus of Nola. But this desire, strong as it may have been, can hardly be held to explain so great a structural change. Let us seek for some architectural reason for the plain wall. Instead of supposing it to be something substituted for the gallery of the basilica when this form was taken over by the Christians, let us imagine it rather *a portion of the older building* left untouched at the time when the columns of the basilica were introduced beneath it. In other words, let us consider the Christian church as the primitive *schola*, with its three plain walls and its apse, enlarged by the introduction of the side aisles and clerestory lighting of the basilica; rather than the pagan basilica, with an apsidal termination added, and blank lateral walls substituted for the convenient and sightly galleries. The non-appearance of these galleries in the early Christian basilica, save in one or two exceptional cases, is a fact requiring more attention than it generally receives, for these were most suitable features in buildings for congregational assembly, and were used almost universally in the circular and domed churches; they reappear also in many Romanesque buildings of the middle ages. That the galleries should have been deliberately abolished for the sake of securing space for pictures is not easy to believe, while the hypothesis that the Christian church begins really in the schola supplies a ready and satisfactory explanation.[1]

[1] O. Mothes, in his elaborate and scholarly work, *Die Baukunst des Mittelalters in Italien*, erster Theil, p. 149, holds that the existing galleries in basilicas, such as those in S. Lorenzo and S. Agnese near Rome, are later additions to the original structures.

The meeting-place of the fourth century is accordingly *not a basilica simplified, but a schola enlarged*, and this fact goes far to explain its form and character. As the union of two different buildings,—a plain walled interior, and a structure put together out of classical colonnades,—it naturally presents a somewhat crude and formless aspect. It is true that it was an admirably convenient building, exactly suited to the needs of the Christian congregations of the time, but in point of architectural character it was really only a building in embryo, and lacked that organic relation of part to part, and of the parts to the whole, which had belonged to the classical temple, and reappeared in the mediæval minster. The materials of the fabric are generally poor, the construction slight. The arrangement of the ground-plan, and the relative proportions of the parts are not fixed on any regular principle. The relation between the width of the nave and that of the side aisles varies in different examples, and the transept, or the clear space before the apse, is sometimes present and sometimes wanting, while the elevation of the side walls of the nave is open to grave criticism. The exterior of the building, lacking as it does those calculated divisions, and that balance of masses which the mediæval architects were so careful to secure, simply follows the form given by the arrangement of the interior, and makes no independent claim on our attention. Hence, in spite of the grand interior effect of the apse of the Christian basilica; in spite of its spaciousness and its simple directness of construction; in spite, too, of the monumental grandeur imparted by its interior decoration in nobly-designed and imperishable mosaics, it fails to produce the impression that it is an artistic unity, informed throughout, like all great architecture, by an all-dominating idea.

Herein resides that difference between the early basilica

and the Romanesque church which we have seen reason to connect with the spirit of the Church at large. In the Rhineland minster of the eleventh and twelfth centuries we find precisely those qualities which are wanting to the basilica. We find, that is, a building remarkable for its well-considered plan, and for the close organic connection of part with part which is observable both in its exterior and interior aspects. The ground-plan is cruciform; the transept is as well marked a feature as the nave itself, and the apse is thrown back by a prolongation of the nave beyond the transept, which makes the fourth arm of the cross. The dimensions of the four arms of the cross are regulated according to a common measure, the unit of measurement being the square space where nave and transept cross each other. Starting with this, the Romanesque architect set off similar square spaces for the chancel and for the two arms of the transept, while the same square, three or four times repeated, formed the length of the nave. The result was a plan like that given in Fig. 26, where a repetition of the square at the meeting of nave and transept generates the whole, the side aisles being exactly half the width of the nave, and divided into squares one-fourth the size of the larger.

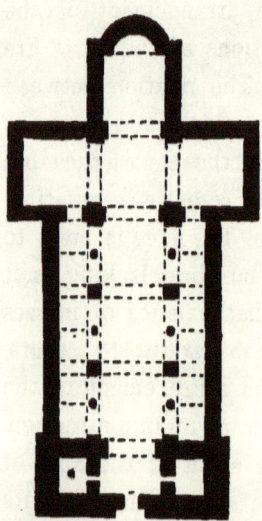

Fig. 26.—Plan of a Romanesque church, simplified from that of Hecklingen.

For the use of columns in the basilica is substituted an entire or partial employment of pillars, furnishing a more suitable support for the upper walls of the nave, while in the completest form of the church, when it is vaulted with stone, connection of the closest kind is established between the different stages of the structure from the floor to the roof.[1] The exterior of the building assumes an equal architectural importance with the interior. The addition of towers supplies a fresh feature, and in their design and grouping the genius of the mediæval architect finds a noble field of exercise. The entrance end of the church, flanked by towers, becomes an imposing and well-proportioned façade, and the terminations of the transept lend themselves to similar treatment. Throughout the building we perceive the working out of an architectural conception, through which the various members are bound together into a grand unity, and just as the early Christian communities, with their loose organisation, give place to the finished Church-system of the middle ages, so from the formlessness of the basilica is evolved the intelligent and beautiful order of a work of art.

It remains for us now to review in a few sentences the chief points of difference between these contrasted buildings, with the object of showing that the appearance of each new feature is not so much the result of Teutonic influence as of a regular and normal architectural development independent of any ethnological changes. The mediæval monasteries were homes of the arts; the buildings connected with them were the work of men who had artistic instincts and training, and it was natural that their designs should show a progressive increase both in expressive character and architectural beauty. The builders of the Syrian churches, so frequently mentioned in these pages, were also men of architectural genius, and they too, as de Vogüé has shown, developed the crude form of the early basilica into structures

[1] It is when stone vaulting came to be used throughout the Romanesque church that the full constructive importance of these square divisions became manifest, for, until the beginning of the Gothic period, the cross-vault employed in all Romanesque districts except southern France was always constructed in the ancient Roman fashion on a square plan.

worthy of comparison with the Romanesque minster. The resemblance of the Syrian style to that prevailing in the South of France has led some writers to derive French Romanesque from Syrian through the channels of communication opened by the Crusades. It seems simpler to regard Syrian and Western Romanesque as parallel developments exhibiting the forms that would naturally be assumed by the basilica among any people of architectural tastes and education.

The cruciform ground-plan of the mediæval church is produced first by the transept, and next by the prolongation of the nave beyond the crossing. The pagan basilica had no transept, nor any vestige of one, but the transept is clearly marked in some of the grander Christian basilicas of Rome, such as Old St. Peter's and St. Paul's, and its introduction may be due to the desire to provide ample space for great ecclesiastical ceremonies. The object of the prolongation of the nave is generally held to be the provision of increased space for the clergy, and the need for this extra accommodation would be especially felt in monastic churches. It is noteworthy that plate 19 of de Vogüé's work shows us an example of the same prolongation of the nave in a church of central Syria. With these extensions the cruciform plan would be arrived at independent of all considerations of symbolism. That the cruciform plan was however known in the earlier period, and that its significance was understood, is proved by the fact that a church of this form was erected by Constantine (*ante*, p. 176), while of S. Croce at Ravenna, built by Galla Placidia in the first half of the fifth century, it is recorded that it was a church *in honorem sanctæ crucis, a qua habet et nomen et formam*. It would seem that St. Ambrose of Milan (374-397) had a special affection for cruciform plans, and many of the churches founded by him were of this form.[1] The basilica proper did not however

---

[1] Mothes, *Baukunst etc.*, p. 140.

assume its cruciform shape till the later mediæval period, for the plans of the age of Charles the Great[1] show it still in process of formation.

With regard to the treatment of the interior surface of the walls of the nave, the change is that from a blank wall between colonnade and roof, to an artistically treated elevation broken into justly balanced horizontal and vertical divisions, with the supporting members of the roof carried down to the ground, and the weak colonnade replaced by massive pillars of the same solid character as the wall itself. The first attempt in the direction of uniting the different portions of this elevation so crudely treated in early Christian basilicas, is to be found in the substitution of round arches in the nave arcade, in place of the flat classical architrave found in some of the great Roman churches. The round arch carries the eye naturally upwards from the column to the wall, and prepares the way for a further union of the parts. When the pillar, a portion of the wall carried down to the ground, is substituted for the column, there is a decided advance. It was natural for the column to be largely employed in Italy, where marble was plentiful and where excellently wrought shafts and capitals could be torn from the ruins of classical edifices. Elsewhere the pillar is almost universal in mediæval churches, and it is sometimes used alone, and sometimes in rhythmical interchange with the column, as in the beautiful St. Michael's at Hildesheim. One of the earliest examples of the use of the pillar in connection with a basilica may be found in the small early Christian church of S. Sinforosa, near Rome, figured in plan upon a previous page.[2]

An important step is taken when the roof story is con-

---

[1] Especially the famous plan of St. Gall, A.D. 822, the discussion of which belongs to the subject of monastic rather than early Christian architecture. [2] *Ante*, p. 65.

nected with the ground by means of a series of half-columns dividing the elevation into vertical sections. Such half-columns find their true function when they act as supports at the springing of the cross-vaults, but they appear occasionally at an earlier period to support arches thrown at intervals across the nave, as at S. Prassede at Rome, or to offer on their capitals a convenient bed for the tie-beams of the wooden roof. On the position of these depends that of the clerestory windows, and of the gallery openings where a gallery exists, while with the orderly spacing of the windows there goes hand in hand a division of the external face of the wall by pilasters of small projection joined under the cornice by a series of round arches. Such an architectural embellishment of the outer wall occurs in early Christian buildings like the fifth century mausoleum of Galla Placidia at Ravenna, but it is in connection with stone vaulting that it receives its full importance. The pilaster is then enlarged into the buttress, which corresponds in position with the half-column within, now charged with important functions as a vaulting-shaft. The roof of masonry renders at once necessary the strictest adherence to plan throughout the edifice, and binds all its parts closely together. It is the true form-giving element in later mediæval architecture, but it is certainly not a Teutonic invention. The fact that, as we have seen, vaults of all kinds were employed from the first in connection with polygonal and domed churches is a sufficient proof of this, and it is significant that the first attempt at vaulting the nave of a basilica occurs at Rome, where in the church of S. Vincenzo alle tre Fontane there exist the beginnings of a vault of brick, which date according to Mothes[1] from the year 796. The age of vaulted basilicas does not come till the eleventh and twelfth centuries, and by that period the process of *organising* the architectural forms of the church

---

[1] *Baukunst etc.*, p. 111.

had been completed in accordance with those principles of Church government which underlie, as has been shown, the facts of architectural history.

There remain two features of a more distinctly northern or Teutonic character. One is the ornamentation of the buildings and the other the use of towers. The first, distinctly unclassical, is the undoubted production of the barbarian conquerors of the empire, and a word will be said upon this Romanesque ornamentation on a subsequent page. With the tower the case is somewhat different. Here is a feature which above all others gives a character to later mediæval architecture. It is the expressive form through which that architecture proclaims itself to be of another spirit to the architecture of the column and architrave. In classical buildings, as in those of Egypt, *horizontal* lines are dominant, in Gothic *vertical*; and it is in the tower that the vertical line shows itself most markedly predominant over the line of base. If Gothic is to be regarded as the expression of the Teutonic spirit, we should be disposed to see in the tower a distinctively northern form. Yet we look in vain for any proof of a primitive tradition of tower-building belonging to the Teutonic peoples, and as a matter of fact not only do towers, from the first epoch of their introduction, appear in the least Teutonised portion of the West— in Rome itself, but some of the characteristic features of Romanesque towers are to be found among the churches of central Syria. At what date and for what purpose towers were introduced into ecclesiastical architecture are much debated questions, and some would even derive their use from the Mohammedans. They appear at Rome and at Ravenna from the seventh century A.D., and one of the earliest existing specimens is that of S. Apollinare in Classe, near the latter city. Their employment seems soon to have become general, and whether as bell-towers or towers of defence they formed the indispensable accompaniment of a group of ecclesiastical buildings. There is a marked difference, however, between the treatment of the tower north and south of the Alps. The Italians place it as a bell-tower apart from the church, while the Rhenish and Norman architects flank their façades with towers, and make them in this way integral parts of their buildings. There is no finer feature in the mediæval churches of the North than this grouping of the towers on each side of the western gable, and on this much of the fine exterior effect of the building depends. Here, however, the Romanesque architect had been preceded by the earlier builders of Central Syria,[1] where, in façades like those of Tourmanin and of Qalb Louzeh, the entrance portal of the church, surmounted by a boldly designed gallery, is flanked by square towers of small elevation which impart at once a fine architectural character to the façade.

In no case, therefore, is it possible to point to any important structural feature in the early mediæval church which is 'Gothic,' in the sense that it is a direct product of the northern spirit or of northern traditions. In all except ornamental forms the spirit and the tradition are still Roman, though in the Romanesque tower the vertical line begins to show its predominance. Romanesque belongs still to the old elements in mediæval society, the home of which was the monastery, and the new elements had to wait till another epoch before their power could make itself felt in art.

The true turning-point in the architectural history of the middle ages is not marked by the irruption into the Roman Empire of the barbarians of the West, but by the great intellectual and political movements of the twelfth and thirteenth centuries. It is then that we meet with the assertion,—ominous for the peace of Christendom,—of freedom of thought

[1] De Vogüé, *Syrie Centrale*, etc., pl. 124 and 135.

and of life, we witness the rise of the vernacular languages of Europe, we watch the decline of the monasteries and the growth of municipal life in the towns, and lastly the consolidation in the France of Philip Augustus of the first centralised modern kingdom. These movements, which are, in a truer sense perhaps than the Renaissance, the beginning of modern Europe, were accompanied by an outburst of religious enthusiasm which found practical outcome in the Crusades. It was at this epoch that the western mind first began to shake itself free from that spell which the Roman system had cast upon it. The thoughts of men began to move more freely, their sight took a wider range, and the enlargement of ideas which was the most valuable result of the Crusades laid the foundation of a new intellectual and artistic development. It was at this time that Roman traditions of architectural effect were set aside, and buildings received a new æsthetic character. That which was least Roman in Romanesque, the use of upward-tending lines in façade and tower, became of greater and greater importance, and the way was prepared for a new style which, while retaining the ancient plan, should be no longer in its spirit Roman but Christian and Teutonic.

It is not the place here to discuss the religious and political conditions connected with the rise of Gothic, or to describe the structural changes in the buildings of the twelfth century through which the transition from Romanesque to Gothic, from round-arched to pointed architecture, was accomplished. It is sufficient here to notice very briefly some of the chief characteristics of Gothic which justify the position here assigned to it. We may first refer to an opinion about Gothic which rests on the highest authority, but which it is necessary to receive with some modifications. This is the opinion that, after all, the origin of Gothic is only a part of the history of construction. M. Viollet le Duc delights to show how the whole elaborate structure of a Gothic building follows by a kind of logical necessity from the adoption of certain simple constructive forms. According to his demonstration, the first step, from which all others followed in regular order, consisted simply in a change in the setting out of the ancient cross-vault on a square plan of the Roman and Romanesque builders. So soon as the diagonal arches of this vault were constructed as half-circles instead of the original weak elliptical curves, the adoption of the pointed arch was necessary for the proper completion of the vault; the use of the pointed arch enabled cross-vaults to be constructed on oblong ground-plans, and the whole of the other consequences followed which made the architecture of the Gothic period.

But, though there is thus the closest connection between Romanesque and Gothic, and though mechanical reasons rather than æsthetic choice, led to the first adoption of the principle of pointed architecture, the fact remains equally clear that Gothic is, artistically and ethically speaking, a new creation. Consideration of structure had the first place in the minds of the beginners of Gothic, but such considerations carry us only a short way towards the understanding of what Gothic became in its perfection, as the outcome of the religious zeal and the rising national life of the Europe of the thirteenth century. There is a significance which carries us beyond the domain of architecture in the Gothic temple with its vast interior spaces, its aspiring arches, its pinnacles which cluster round the high pitched roof, its towers and spires which pierce the air. There is more than geometric science in the long perspective of the nave arcades, with the infinite complexity of interlacing lines where chapel and pillar sweep round the outer circuit of the choir. It was not as a matter of structural consequence or of mere

adornment that these mighty churches became peopled by
those sculptured shapes so graceful in their pose, so tender
in expression; guarded by the company of prophets and
kings who line the solemn portals or wait in their ranks
above; watched by the angel forms which cluster in the
gables or stand in the niches of the pinnacles as if they had
newly alighted from the upper air. And finally is there
not a new clear note of poetic expression struck by the
exquisite foliage carving on capital and moulding, which
brings in the fresh spirit of nature from the fields and wood-
lands? This new spirit which finds its embodiment in
Gothic, is the deep and enthusiastic spirit of the North.

If there is one quality more than another which dis-
tinguishes the northern from the classical temper it is a
certain romantic vagueness, which contrasts with the clear-
ness of vision and love of definite form so characteristic of
the Greek and of the Italian. The opposition between
'classical' and 'romantic' is a common topic of criticism,
and is often illustrated by the contrast between the nicely
measured bounds of that Inferno which Dante surveyed
with so calm an eye, and the awful formlessness of Milton's
Hell. Of similar significance is the fact that in the graphic
arts the use of light and shade, and through them of mystery,
as in themselves vehicles of artistic effect, belongs entirely to
the northern schools, while with the Greeks and Italians the
form was ever the first matter, and mystery is never delighted in for its own sake. This romantic side of the
Teutonic genius, with its power of self-abandonment, its
infinite longings, found an outcome in the ideas of chivalry,
and chivalry took substantial form in the Crusades. Gothic
architecture had its birth in the age of the Crusades, and
expresses the temper of that age as perfectly as the Doric
temple represents the Greek ideal. This is the explanation
of that quality in Gothic which has impressed itself so

deeply upon the popular imagination that it has won for it the reputation of the pre-eminently religious style. It is a quality suggestive of aspiration which seems to reside in the upward striving lines of a Gothic edifice and in its pointed arches; an enthusiastic spirit which speaks in its slender clustering shafts, its towering pinnacles, its multitudinous details, and wealth of lovely forms and colours. It is not so much that Gothic appears religious; it is religious only in a secondary sense through its romantic and enthusiastic character. Now this impression about Gothic, though sometimes dismissed as a mere fancy, has a twofold justification. In the first place it is a well-known fact that Gothic arose under conditions of religious excitement which naturally found vent in artistic as well as in other modes of expression. In the second place there was an undoubted tendency in Gothic forms,—if not from the first, yet from a very early period,—towards height and slenderness which would certainly not have been carried to the extremes they reached at Beauvais and Cologne, had they not been felt to possess some special significance.[1]

This significance is apparent to all who compare the effect of Gothic with Romanesque. The cathedral of Le Mans affords a good opportunity for the comparison, possessing as it does a fine Romanesque nave and a beautiful early Gothic choir. At a distance the towering roof of the choir with its forest of flying buttresses contrasts strikingly with the monotonous mass of the far lower nave. In the interior view it is interesting to note how largely the difference of effect is due to the different character of the round and pointed arches. Along the Romanesque arcades of the nave

---

[1] The nave of Chartres of the latter part of the twelfth century is 106 feet high by some 50 feet wide, while the vault of Beauvais before the middle of the thirteenth had received a height of three times its width, and rose 150 feet above the pavement.

the eye travels round each arch in succession, finding no break, and is carried on from end to end of the series in one unbroken course. In the lofty arches of the Gothic choir the eye ascends the curve from the capital to the crown, and finds then a break. The outline is not continuous, the attention is roused, and we are tempted to leave the arch and carry our glance upwards to the triforium, the clerestory, and the vault. The round arch in a word is self-centred, it keeps the eye within its bound; the pointed leads it perpetually away from itself. The one implies regularity and order, the other has an element of the infinite, and evokes and stimulates thought.

Fully as striking as the height and slenderness of Gothic forms and the special quality imparted to them by the pointed arch, is the character of Gothic decoration.

It is enough to place a carved Romanesque capital from Lombardy or the Rhineland beside a specimen of the sculptor's work from Rheims for a proof how essentially distinct in feeling are the two styles. In the first, where figures are introduced they are massive and of short proportions. Foliage is heavy and reminiscent of classical conventionalism, while the sculptor's fancy is for ever ready to run riot amidst quaint animal forms and grotesques, or to busy itself with curious interlacing lines and bands. This latter style of work almost completely disappears with the coming in of Gothic. The monstrous shapes of dragons and demons and composite beast-men, the curls and knots and flourishes, are soon banished utterly and for ever from the interior of Gothic churches. Only on the exterior, especially in the gargoyles, is the old spirit still allowed to have its way, and the quaint monsters still project their ugly faces along the upper cornice and pour the rain-water through their open jaws. The place of the old Romanesque sculpture is now taken by figures in which slender proportions and a tender sweetness of expression replace the massive grandeur of the older work, and by foliage carving in which the traditions of classical treatment are finally abandoned, and the plants and flowers of the country side are reproduced with natural feeling and a delicacy of execution which have an exquisite charm and carry us forward at once to the modern age.

With regard now to the question of Teutonic influence, it must of course be admitted that if we consider merely the *character of the decoration* of early mediæval churches it will appear as if the stamp of the northern genius were clearly set upon Romanesque, while Gothic has worked itself free from this influence. But it must be remembered that this delight in interlacing lines, in animal forms, and in the grotesque and monstrous, though thoroughly characteristic of the northern tribes, belongs to them as barbarians. It is their inheritance from a condition prior to civilisation, and appears in their earliest decorative efforts in metal-work and carving. Such ornamentation seems indeed to have been the common property of all the tribes which pressed into the West subsequently to the settlement of the Hellenic and Latin peoples. We find it alike among Celts, Teutons, and Scandinavians, and it belongs to a time before the distinct national characteristics of these different peoples had made themselves apparent. Preserved as it was down to the middle ages through that conservatism which in matters of ornament is admittedly so potent, this system of decoration has no pretensions to express the genius of the North in any but the most superficial degree. It is not here that the essential character of that genius asserts itself. Destined as it was in time to find artistic expression in the poetry, the music, the design of a Shakespeare, a Beethoven, a Turner, it was capable of something more in architecture than of merely adding ornament of northern type to classical structures. It was capable of realising itself in all its depth and fulness in a style which it should make essentially its own, and this it accomplished in Gothic. For though the Gothic builders retain unchanged the main features of the Romanesque design, the cruciform ground-plan, the nave and aisles, the clerestory, the façade and towers, yet the sentiment of the building throughout is northern and romantic.

This romantic character of Gothic which is the chief point here insisted on, must not however blind us to an equally important quality of an opposite kind which combines with it to give to the style its high artistic value. Just as the 'romantic vagueness' of the northern temper coexists with a manly grasp on reality which is quite as truly its characteristic, so through all the expressiveness, all the splendid exuberance of Gothic, there is preserved a rational quality which imparts to it an iron strength.

We are brought back here to the point from which this discussion started, the twofold aspect of Gothic structures which have partly a constructive, partly an artistic and ethical interest. As we have seen that *construction* alone does not explain Gothic, so now we must recognise that Gothic is not merely *expression*. That aspect of it which appeals to the student of construction is just as integral a part of it as are those aspects which appeal to the poetic sense or the love of beauty. Only by a due appreciation of both can Gothic be really understood. We may dream beneath the towers or the vault of a Gothic cathedral, and as we dream the whole structure may become to us instinct with life. A moving energy within it may seem to struggle to lift the solid masses upward from the earth. A sense of awe may visit us as we gaze down the solemn aisles or upward to the lonely spires. Voices may speak to us from the myriad sculptured forms or whisper in fairy tones amidst the leaves. We may forget the architect's calculation, the master-builder's care, the toil of mason and carpenter, and feel as if pillar and arch and turret were but the expression of a living will, the clothing of a spirit which stirred and breathed through all.

But we may honour the building equally by approaching it in an entirely different mood and by striving with plummet and measuring-rod to master the secrets of its marvellous construction. The interest of the building from this point of view is not less absorbing because it is of a scientific rather than a poetic kind. The structure of a Gothic cathedral is in truth not less compact though it is infinitely more complex than the simple abbey church of the early mediæval period. Approached from the point of view of construction these buildings are the most scientific that have ever been raised. There is hardly a single part of them, except the spires which crown the towers, which is not an integral part of the organism structurally bound up with all the rest. We may take a single section and notice how the stone vault at the summit has conditioned all the architectural forms below it, giving exact place and quantity to each. If we have given the thrust of the vault in a certain point, the various methods of counteracting the thrust by mass of wall or oblique pressure of flying buttress follow by a mathematical certainty.[1] The parts of the structure all hold as closely together as in Romanesque, but the refined instinct of the Gothic builder has taught him how to break up his principal masses, and clothe in rich and varied work the skeleton of the plan which in the earlier edifices is too plain and obvious for the highest artistic effect.

[1] The Gothic builders themselves had a better way than mere calculation. They seem to have had an inborn genius for the solution of structural problems, and balanced pressure and resistance through happy instinct aided by the lessons of experience. It is reserved for a later time to go over their work after them and to show by calculations and diagrams how scientific was their procedure. See M. Viollet le Duc's *Dictionnaire*, passim.

With what exuberant fulness do detail and ornament seem to be lavished on an exterior like that of Rheims, yet how little there is that is really wayward and fanciful. How different for example is it from a piece of Eastern decoration. While oriental fancy loses itself in a maze of exquisite forms and colours which play aimlessly over the whole field of the design, here every ornament has its nicely calculated place and every detail its due proportion to the larger masses, while the delicate richness of certain parts is enhanced by the simply treated spaces with which they are contrasted. How multitudinous are the slender arrowy shafts which stand sheaved together in the pillars, how intricate seem the lines of the vaulting-ribs round the circuit of the choir, how like the reeds of the river bank do the pinnacles seem to rise about the roof as if each one grew where it listed! Yet we know that every shaft has its appointed work which is set for it to do from the moment that it leaves the floor, and we can see the clustered pillar giving off as it rises towards the roof on this side and on that the single members that encircle the wall-arches or undergird the vault. Geometrically precise are the plans of the vaulting, while the pinnacles have all their predetermined functions in the great organism, lending stability by their weight to the piers from which project the flying buttresses so essential to the safety of the roof. In this way necessity controls the forms which seem so free; and where some would see only poetic expressiveness others discern the successful solution of a technical problem in construction.

The story of Christian architecture in the West thus rounds itself off in the three stages of early Christian, Romanesque, and Gothic. The first was a tentative stage and culminated in the basilica which, though based on Roman models, is an independent creation of the early Church. Fine as it is in its one great interior effect, it has lost something of that architectural consistency for which antique builders had so fine a feeling. This is recovered in the Romanesque church, which is completely Roman in style, and possesses a well wrought-out plan consistent in every part, while the northern spirit reveals itself in the decoration of the buildings with ornamentation of a crude and primitive kind, barbarian rather than distinctively Teutonic. Finally in Gothic the Teutonic temper finds full and free expression, though it is the Teutonic temper barbarian no longer, but disciplined through long ages on the one side by the Church, and on the other by the laws and the institutions of Rome. Gothic retains therefore all that was best in Romanesque, but breathes a new life into the ancient fabric. The solid masses are broken up, the dumb forms become articulate; an ardent aspiring spirit lifts the whole into higher regions of artistic and ethic expressions; and Gothic becomes the first and most perfect embodiment of the genius of the North.

APPENDIX.

NOTE I.—THE PAGAN BASILICA AND THE CHRISTIAN CHURCH.

THE controversy about the origin of the Christian church and the relation to it of the Pagan basilica is of comparatively modern date. Antiquarian interest in the buildings of the past dates from the time of the revival of letters, and we owe to the pen of one of the famous writers of that period the first modern treatise on the Pagan basilica, and indirectly, the first theory of the origin of Christian architecture. The Florentine Leon Battista Alberti, in his work *de Re Ædificatoria*, Florentiis, 1485, describes the basilica in such a manner as to make it clear that he believed it closely to resemble the early Christian church. 'It is apparent,' he writes (lib. vii. c. 14), 'that the basilica was originally a place where the rulers came together under cover of a roof to decide causes at law.' To give dignity to the locality a tribunal, he continues, was added of apsidal form. Then there was a place for promenade (*ambulatio*—the nave), and porticoes, first single and then double. There was also added a second *ambulatio* at right angles to the first (a transept), for the use of the lawyers, so that by joining the two *ambulationes* a T shaped figure was formed. The whole evidently presented itself to his mind in the form of old St. Peter's or St. Paul's at Rome. Nor is this to be wondered at. It was known that the name 'basilica,' which remained still in use for Christian churches, had been also the name of the ancient Roman courts of law and halls of exchange. Enough could be learned about these from Vitruvius and other sources to make it clear that they had, like Christian churches, an oblong form, and were divided in a similar manner into a nave and side aisles. Through a pardonable error, the ancient basilicas were regarded almost exclusively as law courts, and the position of the prætor with his assessors was supposed to correspond to that occupied in the church by the bishop and the clergy. Accordingly the conjecture lay near that the Christian was a copy of the Pagan basilica, and even that the Christian *was* the Pagan basilica, turned from its secular uses to the service of the church. Nay, further, in an age familiar with the 'donation of Constantine' it would seem natural to suggest that this devoted patron of the Church may have handed over Pagan basilicas for the use of the Christians, and we arrive in this manner at an intelligible 'natural history' of the view which makes the basilica the sole and sufficient origin of Christian architecture. Ciampini, in his *Vetera Monimenta*, published in 1690, expresses this view as follows (vol. i. p. 7):—'After the pattern of the ancient basilicas many sacred buildings were erected by the Christians, and retain to this day the name of basilicas. Many also of these old basilicas were turned to sacred uses, and dedicated to divine service,'—and the writers of the last generation received it as an established tradition which we find formulated in the following words by Augusti, *Beiträge zur christlichen Kunstgeschichte*, etc., 1841, i. 22 : '*findet man auch dass solche Gebäude, besonders im IV. Jahrhundert, zur Aushülfe für die neue Staats-religion, zu gottesdienstlichen Versammlungen gebraucht wurden.*'

Precisely the same view has recently been expressed in the articles 'Basilica' in the new edition of the *Encyclopædia Britannica*, and the *Dictionary of Christian Antiquities*, and there is no hint in either that it has ever been seriously called in question. Yet it is the fact that nearly forty years ago, this theory was attacked in a learned and masterly treatise by Zestermann (*die Antiken und die Christlichen Basiliken*, Leipzig, 1847), since which time the old view has only been held, if held at all, in a very modified form; while a voluminous controversy has been carried on, chiefly in Germany, on the question what was, as a fact, the true origin or origins of the Christian meeting-house. It would be a waste of time to give any account of a controversy the very existence of which is completely ignored in the authoritative British publications just mentioned, and it is enough to say that the origin of the church has been sought by one writer or another in the classical temple, the Jewish temple, the synagogue, the crypt and memorial cella of the cemeteries, the private hall, and even the ordinary Atrium, Alæ and Tablinum of the Roman house. The controversy has proceeded till almost all the familiar buildings of the ancient world have been pressed into the service of the theorist, while an incessant process of refutation has resulted in the demolition of every hypothesis as soon as it appeared. It remained to be seen whether a building, unfamiliar and scarcely represented at all in existing monuments, would not after all be found to contain the secret of the beginnings of Christian architecture, and it is the suggestion of the present writer that this building, the pagan *schola*, furnishes the required solution of this interesting problem. That an hypothesis of this kind is the natural outcome of the present state of knowledge on the subject is proved by the fact that in the spring of this year, when these sheets were already in the hands of the publisher, there appeared an elaborate work by Dr. Konrad Lange, of the University of Halle, in which is enforced the same view of the importance of the *schola*—a view which no previous writer has brought into prominence. This work, *Haus und Halle*, Leipzig, 1885, is in many respects a most excellent one. The author insists strongly on the suitability for Christian use of the *schola*, with its oblong plan, and its apse for the presiding officials, and finds in it the most probable earliest form of the church, though he is unable to offer any absolute proof that such *schola* were used by the Christians. It will be interesting to see whether the direction now given to investigation on this subject will lead to the discovery of the desired evidence. The main portion of Dr. Lange's book is, however, devoted to the basilica, and he treats this in a broad manner, in connection with the columned halls of the ancient world in general, bringing together in convenient form a mass of most valuable literary material. The work concludes, however, with what the present writer cannot but consider a most unfortunate attempt to rehabilitate the old theory of the acquisition of the Pagan basilicas by the Christians of the age of Constantine.

This is a point of such capital importance in the early history of Christian Architecture that the following criticism, both of the theory in general and the use made of it by Dr. Lange, may not be unwelcome to the student.

The theory has, it will be noticed, two forms, according as we suppose the Christians to have *taken possession of* or merely *imitated* the Pagan basilicas. The matter is expressed thus in the *Dictionary of Christian Antiquities*, art. ' Basilica ':—' When Christianity became the religion of the state, these buildings ' (basilicas) ' were found to be so well adapted to the celebration of public worship that some were by some slight modifications fitted and used for the purpose, and the new buildings constructed expressly to serve as churches were built almost universally on the same model.' Let us examine these two hypotheses in succession.

Consider in the first place what this handing over of public basilicas to the Christians would imply. The buildings in question combined the functions of exchanges and law-courts.

Always in close proximity to the Forum, the busy centre of the life of an ancient city, they were among the most conspicuous and important of its structures. Is it likely that at a time when the Empire was still flourishing they would be suddenly withdrawn from secular use and handed over to the members of a religious association? When the Roman empire in the West had fallen and the ancient life of its cities was over, then it was natural that the Church, the most powerful organisation of the time, should appropriate disused public buildings both sacred and secular. But neither commerce nor jurisprudence declined when Christianity became a religion recognised by the law. Dr. Lange (*Haus und Hulle*, p. 313) seeks, it is true, to combat this objection by suggesting that Rome, at any rate, was in rapid decline owing to the growing importance of Constantinople. But Constantinople was not dedicated till nearly twenty years after the toleration edict of Constantine, and St. Chrysostom implies that up to his time, at the end of the fourth century, it was not more important than Antioch (*de Statuis*, hom. iii. 1). Rome cannot have seriously declined till the barbarian invasions of the beginning of the fifth century. Hence we should need strong evidence for the assertion that any large number of exchanges and courts of law were in consequence of the edict of Milan turned from their ordinary purposes. Let us however suppose for the moment that this was actually the case, and note what follows. Constantine's famous edict of Milan in favour of the Christian Church was only an edict of universal religious toleration, and it was not till a subsequent period that Christianity became the established religion of the Empire. We are to imagine accordingly the adherents of this lately persecuted but now tolerated sect turning the lawyers, clients, merchants, and brokers out into the streets, and taking exclusive possession of some of the largest and grandest buildings of the cities. What a triumph in a worldly sense for the Church! What an exaltation of the recently despised sectaries! What an occasion for pride and glorification to the numerous ecclesiastical writers who are never weary of exalting the majesty of the Church and the merits of its imperial and princely patrons! The laudation of Constantine in particular was carried to extravagant lengths by Eusebius, and other patristic writers, and here was a theme on which they might be expected to dilate with fullest satisfaction. When we turn, however, to these panegyrists, what is it that we find? Will it be credited that there is not a single direct notice in any one of them of this supposed donation of Constantine to the Christians? In no single instance is there any mention that either this Emperor or his successors handed over any one public building of the kind for such a purpose. This important gift, if it was ever offered, was never even acknowledged by the recipients. Now the absolute silence which reigns on this point is a convincing proof that the supposed donation of the basilicas is a mere fiction of later times. Against negative evidence such as this no theory can really hold its ground, even though it were supported by substantial secondary evidence. In this case there is no such substantial evidence forthcoming, and that which is made to do duty for it may be judged of from the following instance. In a certain rhetorical composition by Ausonius, containing the writer's thanks to the Emperor Gratian for the bestowal of the consulship, occur the following words: . . . '*basilica olim negotiis plena nunc votis pro tua salute susceptis,*' . . . 'the basilica formerly full of business, now of vows undertaken for thy welfare,' and these have been quoted (see *Ency. Brit.*, art. 'Basilica') to prove that the basilica had been turned from commercial purposes to those of Christian assembly. Whether or not a Christian service of those days can be adequately described as consisting of 'vows for the welfare of an Emperor,' those who rely on this passage do not say; but did they quote the whole sentence of which these words form a part, they would see at once how mistaken is their inference. The following is the passage. The object of Ausonius is to convey the most extravagant expression of gratitude to Gratian, and he accordingly begins by saying that he is thanking him continually at all times and places:—'*usque quaque gratias ago, . . omni loco, actu, habitu et tempore.*' Nay, the whole town rings with laudation of the Emperor:— '*quis, inquam, locus est, qui non beneficiis tuis agitet, inflammet? Nullus, inquam, Imperator auguste, quin admirandam speciem tuæ venerationis incutiat: non palatium, quod tu, cum terribile acceperis, amabile præstitisti: non forum, et basilica olim negotiis plena, nunc votis, votisque pro tua salute susceptis: . . . non curia honorificis modo læta decretis, olim solicitis mæsta querimoniis,*' etc. Unless we are to suppose that the palace, the forum, and the senate-hall had all like the basilica been turned into Christian churches, we must naturally take the passage as implying that in all the public places where men came together there was formerly nothing talked of but business and the dangers of the times, while *now* the praises of the Emperor resounded on every side. With this passage before him it is a *mauvaise plaisanterie* on the part of the writer in the *Encyclopædia* to pick out the words '*basilica* . . . *susceptis*,' and attempt to make them mean something utterly foreign to the idea in the mind of Ausonius.

Dr. Lange seeks to prove the acquisition of public basilicas by the Christians by some almost equally fallacious arguments. '*Für die Ueberlassung von Kaufhallen an die Christen*,' he writes (*Haus und Halle*, p. 313), '*haben wir nun in der That vier Beweise.*' The first of these proofs he finds in a passage in St. Jerome (*Ep.* xxx.), describing an incident as having taken place *in basilica quondam Laterani, qui Cæsariano truncatus est gladio*, etc. This Lateranus was put to death by Nero; he inhabited a mansion celebrated by Juvenal (*Sat.* x. 15) as '*egregiæ Lateranorum ædes*,' and 250 years later this mansion, or its successor, belonged to Fausta, the wife of Constantine, who handed it over to the purposes of Christian assembly. There is nothing here which was not perfectly well known before. What Dr. Lange wishes us to believe is that the 'basilica' of this mansion was not a private hall of audience, like that in the palace of Domitian (*supra*, p. 127), but a sort of public bazaar appended to the house, and so to all intents and purposes a public not a private basilica. His argument for the existence of such semi-public basilicas (*Haus und Halle*, 199, 222) merits attention, but in no case have we the slightest reason to believe that this basilica of the Laterani was anything other than one of the private basilicas mentioned by Vitruvius as forming a part of the palaces of the great. Here are his words (*lib.* vi. c. 8) : '*nobilibus vero qui honores magistratusque gerundo præstare debent officia civibus, faciunda sunt vestibula regalia alta, atria et peristylia amplissima, silvæ ambulationesque laxiores ad decorem majestatis perfectæ, præterea bibliothecæ pinacothecæ basilicæ non dissimili modo quam publicorum operum magnificentia comparatæ, quod in domibus eorum sæpius et publica consilia et privata judicia arbitriaque conficiuntur.*' Of this kind were also the edifices spoken of by St. Jerome (see *Haus und Halle*, p. 314) as '*instar palatii privatorum extructæ basilicæ*,' and especially the basilica of Theophilus at Antioch, mentioned in the *Recognitions* (see *supra*, p. 44). Lange (p. 318) would turn this also into one of his semi-public bazaar-basilicas, but he forgets the earlier passage in the *Recognitions* about the private hall of Maro opening apparently into his garden, and holding 500 people. The 'basilica' of Theophilus was meant, we may be sure, to be a hall of the same character. Another of Dr. Lange's 'Beweise' is based on an inscription of uncertain date, recording the death of an infant, '*in cujus honorem basilica hæc a parentibus adquisita contectaque est.*' How utterly improbable it is that the parents of this two-years-old infant should, as Dr. Lange supposes, have sought to do him honour by purchasing and repairing a disused and ruinous public or semi-public basilica! The explanation of the passage is that the word 'basilica' soon came to be very loosely applied, and one of its later uses was equivalent to 'funeral monument,' which is the natural sense for it in the inscription.

There remains now but one of the proofs brought forward by Dr. Lange, and this is a somewhat famous passage describing an incident which took place in the year 366, '*in basilica Sicinini, ubi ritus Christiani est conventiculum*' (Amm. Marcel. xxvii. 3, 13, Lange, p. 315). This passage has been variously interpreted, and most writers seem agreed that the '*basilica Sicinini*' was, like the '*basilica Laterani*,' part of a private palace. Dr. Lange makes it a public basilica, and we are willing to agree with him, for, upon his own showing, the matter furnishes the clearest proof of the truth of the view of the basilica taken in these chapters. Another ancient writer dealing with this same incident (Sokrates, *Hist. Eccl.* iv. 29) speaks of certain Christians as meeting at the time ἐν ἀποκρύφῳ τόπῳ τῆς Βασιλικῆς τῆς ἐπικαλουμένης Σικίνης 'in an out-of-the-way corner of the Sicinian Basilica,' and Dr. Lange takes this as implying that '*die "basilica Sicinini" war eine profane Basilika, in der ... ein christliches " Conventiculum" befand. Sie war also selbst damals noch keineswegs ganz dem christlichen Cultus überliefert, sondern nur ein Theil derselben, vielleicht eine Ecke, wurde zu gottesdienstlichen Versammlungen benutzt.*' ... Now this meeting of the Christians in a part of a public basilica—if such was the 'basilica Sicinini'—agrees exactly with our general view of this building. It was not as a whole 'well adapted for the celebration of public worship,' but for the transaction of various kinds of business at once, and there is no harm in believing that the Christians might choose a certain part for their gatherings, perhaps railing it off from the rest. This no more implies a formal surrender to the Christians of the whole building, than the words in Acts v. 12, 'they were all with one accord in Solomon's Porch,' imply that a portion of the Temple courts had been handed over by the Sanhedrin to the followers of Jesus of Nazareth.

We may take it therefore as proved, that there is no evidence of any kind, that public or semi-public bazaar basilicas were, in the fourth century, handed over for the purposes of Christian worship.

On the other form of the hypothesis under discussion, that which concerns the *imitation* by the Christians of the basilica, little need here be added to what has already been laid before the reader. The whole question turns upon the existence and the position of the apse in the pagan basilica; those who adopt the old view must either show that the apse figured in both basilicas alike, or that the apse and its position are comparatively unimportant matters, on which the resemblance or difference of the buildings does not really depend. The treatment of this point by Bunsen is instructive. Bunsen's authority went far to establish the hypothesis we are considering as the orthodox view of the beginnings of Christian architecture, and his fine description of an ancient basilica, in his *Basiliken des christlichen Roms*, München, 1842, has given that building its standard form in the popular imagination. Like the writers of the Renaissance, Bunsen imagined the basilica as pre-eminently a *law court*. It had always an apse '*das edelsten Theil des Gebäudes*' (p. 19), and before the apse a free space like that in a large Christian basilica. Here then occurred a difficulty. Bunsen admits that the colonnades probably ran across the small end of the building and carried galleries, from which spectators could watch the proceedings before the tribunal in the apse (p. 20), yet at the same time he appears to imagine that the apse would still dominate the interior, as in the Christian church. On entering the building, he says, '*so haben wir eine architektonische Einheit vor uns . . . Die erhöhte Tribunalsnische ragte dem Eintretenden als die Spitze des ganzen entgegen und das Tribunal, der Unterbau des Richtersitzes, mit diesem selbst, fesselte sein Auge*' (p. 22). We have only to turn back to Fig. 19 to see what the real fact of the case would be. What sort of an effect would even the grandest apse produce when seen through a screen composed of a double colonnade, with its heavy entablature, surmounted by a second double row of columns in the gallery? It is obvious that the apse would form no part of the architectural effect at all, and one reason from which we may conclude that the apse did not belong to the architecture of the pagan basilica, is the fact of the awkward effect it would produce, half-seen, half-hidden, beyond the colonnades and gallery.

Again, a glance at Figs. 19 and 20 is enough to show the fallacy of the view to which Dr. Lange gives expression, that it did not really matter, from the architectural stand-point, whether or not there was an end colonnade before the apse. To leave this out, he says (*Haus und Halle*, p. 310) 'would be an unimportant innovation'! It is inconceivable that any one with an idea of what architecture means could maintain this, or fail to see that the position of the apse is really the *essential element* in the architectural effect of the early Christian church, while the continuous colonnades are just as essential to the character of the ancient hall of exchange.

The only hope for the basilican theory would be in proving that the pagan basilica showed examples of this free position of the apse. It is true that the basilica of Maxentius (or Constantine) had an open nave terminated by an apse, but this building, as we have noticed, was not a basilica proper. It was a vaulted not a columned structure, and was imitated from the halls of the Thermæ. Its date, early in the fourth century, makes it quite possible that the terminal apse was really due to the influence of Christian architecture, which at that period had probably assumed its distinct characteristics. In the case of the columned basilicas, with which alone we are concerned, we cannot make a complete list of examples, for there are remains of many buildings which *may* have been basilicas, but are equally likely to have been halls of Thermæ, *scholæ* or Christian churches. An exhaustive series of basilicas is therefore impossible, but of tolerably certain examples there is only one, the basilica at Otricoli, now entirely destroyed, in which we seem to have the desired instance. There was here a square interior divided by columns into nave and aisles. At the end of the nave was a spacious apse, but we are expressly told that in front of the apse were discovered the remains of two columns the same height as those of the main colonnades (*Haus und Halle*, p. 233). It is clear therefore, that this building should be restored with a *masked* not a *commanding* apse. The whole matter cannot be better summed up than in the words of Dr. Lange himself (p. 362), 'ist die Herumführung der Seitenschiffe auch an den Schmalseiten bei Basiliken, wie wir gesehen haben, überhaupt typisch,' and with the establishment of these end colonnades as typical, is involved the fall of the ancient theory of an essential similarity between the pagan and the Christian basilicas.

Note II.—Sassanid Architecture.

No account of the history of dome construction can be complete without a reference to the domed palaces of Persia, generally ascribed to the period of the Sassanid monarchs, contemporaries and rivals of the early Byzantine emperors. The Persian Sassanidæ, who ruled from 226 to 651 A.D., belonged to a national dynasty representing a reaction from the influence of Greek civilisation, which had been all-powerful under the preceding régime of the Parthians. Disposing of the immense resources of the nearer East, the greatest of the Sassanidæ,—Ardeshir, the Sapors, the Khosroes,—waged equal and often victorious wars with the representatives of Rome, and Persia became under their rule the second power in the world. Not less famous were some of these monarchs, notably the first Khosroes surnamed Nishurwan, for the arts of peace, and it would be natural to expect some abiding architectural monuments of their greatness. Such monuments exist in the form of sundry ruined palaces, such as the so-called 'throne of Khosroes' in that monarch's capital of Ktesiphon on the Tigris, and the connection of these with the Sassanid rule has been an accepted fact in architectural history. The standard work on these buildings is that by Flandin and Coste, *Voyage en Perse*, published about 1845, and in that are figured and described the remains of Sassanid buildings and of Sassanid sculptured monuments of the same period. The distinguishing characteristic of this architecture is a large employment of the arch and of the dome in forms which at once invite comparison with those of Rome and Byzantium. Thus the chief feature of the 'throne of Khosroes' is a grand arch 70 feet in width and 85 in height forming the entrance hall of the palace. The ruined piles of Firuz-Abad and Sarbistan, in the district of Fars, the heart of ancient Persia, possess circular cupolas of wide span and of elliptical section, united by means of pendentives to a square base. With these buildings are naturally connected interesting questions concerning the history of the pendentive and the relation of Sassanid to Byzantine domes. It may be inquired too how far these arches and cupolas represent an outcome of old traditions of vault construction in the East, how far they can be affiliated to the architectural forms of the

Romans. All these are fair matters for discussion and leave untouched the central fact that Sassanid domes date at any rate three or four centuries later than the Pantheon of Agrippa. Quite recently however a French engineer and explorer, M. Marcel Dieulafoy, in a work entitled *L'Art antique de la Perse*, has propounded the startling theory that two at least of these monuments, the palaces of Firuz-Abad and Sarbistan, are not Sassanid works at all, but belong to the ancient Achemenid empire of Darius and Xerxes. Such a theory, which if found to be true would alter entirely our views upon the history of the dome as a monumental feature, needs strong evidence for its support, and when the author in his preface speaks of *les monuments voûtés du Fars, qui appartiennent, je l'annonce immédiatement, à la période achéménide*, we naturally turn to the part of the work devoted to these monuments with the most lively curiosity. It is a disappointment to find that so far as the book has proceeded there is no attempt made to bring forward any new evidence upon the subject. M. Dieulafoy offers, it is true, most valuable technical observations on these interesting monuments, and accompanies them with excellent photographic and line illustrations, but for any proof of this altogether novel view that Firuz-Abad and Sarbistan are not really Sassanid buildings we look in vain. Nor is the received view ever seriously combated. The following, for example, are the remarks of Flandin and Coste (*Voyage en Perse, Perse ancienne*, p. 176) on the resemblance of the 'throne of Khosroes' to the palace of Firuz-Abad:—'*ce monument par sa disposition et le genre de sa décoration extérieure rappelle le palais de Firouz-Abâd. S'il y a entre eux quelques légères différences de style et de caractère, elles ne sont pas assez sensibles pour qu'on ne les rapporte pas tous deux à une ère commune qui est évidemment celle des Sassanides.*' That the 'throne of Khosroes' cannot belong to the ancient Persian empire is proved by the fact that the city of Ktesiphon wherein it lies was not founded till the Parthian era; and, this being the case, substantial differences must be pointed out between it and Firuz-Abad before the latter can be taken from the place assigned to it by Flandin and Coste, and accepted as the senior of the other by five hundred years. Yet M. Dieulafoy only brings forward one single feature about Firuz-Abad which makes the monument as belonging, in his opinion, to the early period. This is the treatment of the doors, which are set in a square frame and surmounted above by the familiar Egyptian cornice. Doorways of this form are conspicuous among the remains of the ancient Persian palaces at Persepolis, and

from this fact M. Dieulafoy immediately concludes that Firuz-Abad is of the same date as the palaces of Darius and Xerxes. '*La decoration des portes de Firouz-Abâd devrait, à elle seule, dater le monument. Il semblerait, en effet, que chaque baie portât gravé sur ses linteaux, son couronnement et ses profiles, le sceau des Achéménides*' (*L'Art antique de la Perse*, 4me partie, p. 59). For those who would answer that the Sassanid builders of Firuz-Abad simply copied the famous old doorways of Persepolis, M. Dieulafoy has a rejoinder ready:—'*jamais . . . les Parthes ou les Sassanides n'auraient eu l'idée fort moderne et fort singulière de faire des mauvaises restitutions archéologiques et d'aller chercher dans les ruines de monuments vieux de cinq à six cents ans, pour les ajuster à un édifice bâti à la mode du jour, des modèles de portes.*' (*Ibid. l.c.*). But was not the Sassanid empire itself a 'restitution archéologique,' a revival of the glories of Cyrus and his successors? If it be true that the founder of the Sassanid dynasty demanded from the Romans the cession of all the provinces which had once formed a part of the domains of Cyrus and Darius, is it not perfectly natural that he or one of his successors should imitate in his palace some architectural features belonging to the glorious period which he burned to restore? It is curious, moreover, that these very doorways afford some rather strong evidence that they were borrowed in part from the architecture of the Romans. Within the post-and-lintel framing, the actual doorway is formed by a round arch springing from upright piers crowned by moulded imposts. Such arches, especially in combination with the rectangular framing, are characteristic Roman forms, and would in themselves go far to prove the Sassanid origin of the whole palace, were that ever seriously called in question. We cannot in conclusion regard M. Dieulafoy's treatment of this subject as in any way correspondent to its importance: nor is it without something akin to amazement that we read the following in M. Choisy's recent work, *L'Art de bâtir chez les Byzantins*:—
'*M. Dieulafoy établit, par des rapprochements que la photographie rend saisissants, que l'architecture de Servistan et de Firouz-Abad est contemporaine de celle de Persépolis, et que la civilisation sassanide n'a rien à revendiquer en elle*' (p. 154, note). Whether this view is true or not we may be content to leave an open question, to be studied by those who have the necessary opportunities for interrogating the monuments themselves. That the view is in any sense of the word '*établi*' by anything which occurs in the first four parts of M. Dieulafoy's work we must beg emphatically to deny.

INDEX.

AACHEN, Minster at, 151, 191.
Agape, 14, 25 n., 59 n.
S. Agnese, Rome, 64, 198 n.
Alexandria, 83, 94.
Altar, Christian, 7, 120.
S. Andrea in Barbara, Rome, 62.
Antioch, earliest domed church at, 148 f.
S. Antonio, Padua, 151.
S. Apollinare in Classe, Ravenna, 119, 205.
—— Nuovo, Ravenna, 75, 131.
Apse, 106, 119 ff., 197, 224 f.
Arch of Triumph, 119, 131.
Architecture, Ancient, in relation to Christian, 79 ff.
—— Egyptian, 80.
—— Mesopotamian, 80 f.
—— Hellenistic, 81 ff.
—— Sassanid, 140, 226 ff.
—— primitive, of Celts and Teutons, 193 ff.
—— its continuity under Roman and barbarian Rule, 187.
Ark containing the rolls of the Law, 105 ff.
Atrium, 42, 116, 162.

BAPTISTERIES, 74 f., 146.
Baptistery at Ravenna, 76 n., 154.
Basilica, Pagan, 66, 76 ff.
—— Christian, 115 ff.
—— Christian and pagan contrasted, 123 f.
—— true relation of Christian and pagan, 217 ff.
—— compared with Romanesque minster, 199 f.
—— private, 68, 126 f.

Basilicas at Rome, Porcia, 78 f., 123.
—— of Maxentius or Constantine, 87, 225.
—— Ulpia, 91, 122 ff.
—— Julia, 91 f., 122.
—— Laterani, 222.
—— Sicinini, 223.
Basilica at Pompeii, 90 f., 122.
—— at Otricoli, 225.
—— described by Vitruvius, 90.
—— (synagogue) at Alexandria, 101 ff.
—— bazaar, of Dr. Lange, 222.
Basilicas, Christian, at Holy Sepulchre, 116, 128 f.
—— at Tyre, 116.
—— (Ursiana) at Ravenna, 76 n.
Birthdays of the Dead, 17, 24, 31.
Bosrah, Syria, church at, 149.
Burial-Clubs, 17.
Byzantine culture and art, 166 ff.

Cantharus, 116.
Capitoline plan of Rome, 40 f., 52 f.
St. Cæcilia, 27 f.
Catacombs, wall paintings in, 129.
S. Cecilia in Trastevere, Rome, 63.
Cellæ, memorial, 19, 25 f., 56 ff., 120 f.
Cemeteries, Ancient, 21.
—— Christian, 20 ff.
Cenacula, 40 f.
S. Clemente, Rome, 63.
Clerestory, 75 f.
Clubs and Assemblies, 12 f., 16 f., 49.
Columned Architecture, 79 ff.
Common Meals, 14.
Confessio, 66, 120.
SS. Cosma e Damiano, Rome, 130.

S. Costanza, Rome, 146, 157.
S. Croce, Rome, 63.
—— Ravenna, 202.
Crypt, 66.
Curiæ, 50 ff.

DECORATION, Christian, in Catacombs, 129.
—— in Basilicas, 128 ff.
—— Byzantine, 166 ff.
Divisions, interior in churches, 61.
Domes, 135 ff.
—— later use, 176.
—— as a Christian form, 177.
'Dome of the Rock,' 147.
Domed churches, earliest, 148.

EDICT of Valerian, 22.
—— Gallienus, 22.
—— Diocletian, 70.
—— Milan, 22, 220.
—— Constantine, 72.
—— Theodosius, 72.
Ezra, Syria, Church at, 149.

FEAST OF CHARITY, 14.
Festivals of the Martyrs, 24, 31 f.
Fossores, 57 ff.
S. Front, Périgueux, 151.
Funeral Colleges, 17 ff.
—— customs and ceremonies, 17 ff., 26.
—— *Schola*, 53.

HALL, hypostyle, at Karnak, 80, 93 f.
Hellenistic architecture, 81 ff.
—— cities, 83.
Holy Sepulchre, 147.
House, ancient, 39 ff.
Houses, private, used by Christians, 43 ff., 67 f.

Insulæ, 41.

S. JOHN LATERAN, 63.

KHORSABAD, 81.

LE MANS, Cathedral, 210.
S. Lorenzo at Milan, 33, 150, 157 ff.
—— Rome, 64, 198 *n*.
Love-feast, 14 f.

S. MARIA DEGLI ANGELI, Rome, 85.
—— Maggiore, . . ,, 131.
—— in Trastevere, . ,, 64.
—— ad Martyres, . ,, 32.
S. Mark's, Venice, 134, 151, 176.
Mausoleum of Galla Placidia, Ravenna, 134, 176, 204.
—— Theodoric, Ravenna, 191.
Meetings, Christian, in private houses, 38 ff.
Memorial feasts, 17 ff.
—— of the Martyrs, 24 f, 31, 56 ff.
S. Michael's, Hildesheim, 203.
'Minerva Medica,' Rome, 145, 152 f., 156, 160.
Minster, Romanesque, 200 ff.
Monasteries, 45, 182.
Mosaics, Christian, 130 f., 134.
—— at Byzantium, 170 f., 174.

Narthex, 162.
Naukratis, 83 *n*., 94.
Nicomedeia, meeting-place at, 70 f.

Oeci, 42 f.
—— Egyptian, 77, 93 f.
Oratory, private, 47.
Ornamentation, Byzantine, 171 ff.
—— Celto-Teutonic, 173, 212.
—— Romanesque, 211 f.
—— Gothic, 209, 211 f.
Ostia, *scholæ* at, 50.

PANTHEON, Rome, 85, 136, 140 ff., 152, 155 f., 177.
S. Paul's, Rome, 64, 119, 131, 202.
—— London, 177 f.
Pergamon, use of the arch at, 83.
Peristyle, 42 f.
S. Peter's, old, Rome, 64, 118 f., 202.
—— new, ,, 177 f.
Pharisees, 109 ff.
S. Prassede, Rome, 204.
Precinct, of early Christian basilica, 116.
Presbyterium, 7, 120.
Preservation of bodies in Catacombs, 27.
S. Pudentiana, Rome, 63, 131.

QALB LOUZEH, Syria, 206.

RELICS OF MARTYRS, 24, 32, 65.

INDEX.

'Roman manner,' the, 194 f.
Romanesque architecture, 200 ff.

S. SABINA, Rome, 117, 129.
Scholæ, 14, 17, 48 ff., 68, 197 f., 219.
S. Sepolcro, Bologna, 147.
SS. Sergius and Bacchus, Constantinople, 162.
Sexes separated in the churches, 61.
S. Sinforosa, Rome, 64 f., 203.
Sodalicia, 13.
S. Sophia, Constantinople, 160 ff., 175 ff.
στοὰ βασίλειος, 79.
Synagogue, Jewish, 8 f., 34 ff., 65 n., 67, 70 n., 96 ff.
Synagogues of Galilee, 96 f., 99 f.
Syria, Central, churches of, 33, 149, 196, 201 f., 205 f.

TEMPLARS' CHURCHES, 148.
Temple, Jewish, 7.

Temple of Herod, 77, 82.
Tenuiores, 17.
Teutonic invasions, their effect on architecture, 183 ff.
Tourmanin, Syria, 206.
Towers, 205 f.
Triclinium, Christian, 46, 54.
—— at Pompeii, 58.
'Trophies of the Martyrs,' 55, 69.

UPPER CHAMBERS, 1, 38 ff.
—— stories, 40 ff.
Underground meeting-places, 22, 30, 60 ff.

VAULTING, 80 ff., 135 ff., 201 n., 204.
Vestibules, 117.
S. Vincenzo alle tre Fontane, Rome, 204.
S. Vitale, Ravenna, 150, 162.

ZESTERMANN, on Basilicas, 78 n., 218.

15A Castle Street,
Edinburgh, *March* 1886.

LIST OF BOOKS PUBLISHED BY
DAVID DOUGLAS.

On the Philosophy of Kant.
By ROBERT ADAMSON, M.A., Professor of Logic and Mental Philosophy, Owens College; formerly Examiner in Philosophy in the University of Edinburgh. Ex. fcap. 8vo, 6s.

The Age of Lead: A Twenty Years' Retrospect.
In three Fyttes. "VAC VICTIS." Second Edition. Sm. crown 8vo, 2s. 6d.

The Correspondence of Sir Patrick Waus of Barnbarroch, during the latter half of the Sixteenth Century, from originals in the Family Charter-Chest. Edited by R. VANS AGNEW. Demy 8vo, 21s.

Stories by Thomas Bailey Aldrich.
THE QUEEN OF SHEBA. 1s., or in cloth, gilt top, 2s.
MARJORIE DAW, and other Stories. 1s., or in cloth, gilt top, 2s.
PRUDENCE PALFREY. 1s., or in cloth, gilt top, 2s.
THE STILLWATER TRAGEDY. 2 vols. 2s., or in cloth, gilt top, 4s.
FROM PONKAPOG TO PESTH. [*In the Press.*

"Mr. Aldrich is, perhaps, entitled to stand at the head of American humourists."
—*Athenæum.*
"*Prudence Palfrey* is a delightful novel, sweet and wholesome, and with exquisite dashes of humour."—*Nonconformist.*
"*Marjorie Daw* is a clever piece of literary work."—*Saturday Review.*

Johnny Gibb of Gushetneuk in the Parish of Pyketillim,
with Glimpses of Parish Politics about A.D. 1843. By WILLIAM ALEXANDER. Eighth Edition, with Glossary, ex. fcap. 8vo, 2s.
Seventh Edition, with Twenty Lithographic Illustrations—Portraits and Landscapes—by GEORGE REID, R.S.A. Demy 8vo, 10s. 6d.

"A most vigorous and truthful delineation of local character, drawn from a portion of the country where that character is peculiarly worthy of careful study and record."—*The Right Hon. W. E. Gladstone.*

Life among my Ain Folk.
By WILLIAM ALEXANDER.
Contents.
1. Mary Malcolmson's Wee Maggie.
2. Couper Sandy.
3. Francie Herriegerie's Sharger Laddie.
4. Baubie Huie's Bastarl Geet.
5. Glengillodram.
Ex. fcap. 8vo. Second Edition. Cloth, 2s. 6d. Paper, 2s.

"*Baubie Huie's Bastard Geet*, which is full of quiet but effective humour, is the clearest revelation we have ever seen of the feeling in Scotch country districts in regard to certain aspects of morality."—*Spectator.*
"We find it difficult to express the warm feelings of admiration with which we have read the present volume."—*Aberdeen Journal.*

LIST OF BOOKS

Notes and Sketches of Northern Rural Life in the Eighteenth Century, by the Author of "Johnny Gibb of Gushetneuk." Ex. fcap. 8vo, 2s. Cloth, 2s. 6d.

"This delightful little volume. It is a treasure."—*Daily Review.*

David Douglas's "American Authors."

Latest Editions. Revised by the Authors. In 1s. volumes. By Post, 1s. 2d.
Printed by Constable, and published with the sanction of the Authors.

By W. D. HOWELLS.
A FOREGONE CONCLUSION.
A CHANCE ACQUAINTANCE.
THEIR WEDDING JOURNEY.
A COUNTERFEIT PRESENTMENT.
THE LADY OF THE AROOSTOOK. 2 vols.
OUT OF THE QUESTION.
THE UNDISCOVERED COUNTRY. 2 vols.
A FEARFUL RESPONSIBILITY.
VENETIAN LIFE. 2 vols.
ITALIAN JOURNEYS. 2 vols.
THE RISE OF SILAS LAPHAM. 2 vols.

By FRANK R. STOCKTON.
RUDDER GRANGE.
THE LADY OR THE TIGER?

By GEO. W. CURTIS.
PRUE AND I.

By J. C. HARRIS.
(*Uncle Remus.*)
MINGO, AND OTHER SKETCHES.

By GEO. W. CABLE.
OLD CREOLE DAYS.

By B. W. HOWARD.
ONE SUMMER.

By JOHN BURROUGHS.
WINTER SUNSHINE.
PEPACTON.
LOCUSTS AND WILD HONEY.
WAKE-ROBIN.
BIRDS AND POETS.
FRESH FIELDS.

By OLIVER WENDELL HOLMES.
THE AUTOCRAT OF THE BREAKFAST TABLE. 2 vols.
THE POET. 2 vols.
THE PROFESSOR. 2 vols.

By G. P. LATHROP.
AN ECHO OF PASSION.

By R. G. WHITE.
MR. WASHINGTON ADAMS.

By T. B. ALDRICH.
THE QUEEN OF SHEBA.
MARJORIE DAW.
PRUDENCE PALFREY.
THE STILLWATER TRAGEDY. 2 vols.

By B. MATTHEWS and H. C. BUNNER.
IN PARTNERSHIP.

*** *Other Volumes of this attractive Series in preparation.*

Any of the above may be had bound in Cloth extra, at 2s. each vol.

"A set of charming little books."—*Blackwood's Magazine.*
"A remarkably pretty series."—*Saturday Review.*
"These neat and minute volumes are creditable alike to printer and publisher."—*Pall Mall Gazette.*
"The most graceful and delicious little volumes with which we are acquainted."—*Freeman.*
"Soundly and tastefully bound . . . a little model of typography . . . and the contents are worthy of the dress."—*St. James's Gazette.*
"The delightful shilling series of 'American Authors' introduced by Mr. David Douglas has afforded pleasure to thousands of persons."—*Figaro.*
"The type is delightfully legible, and the page is pleasant for the eye to rest upon, even in these days of cheap editions we have seen nothing that has pleased us so well."—*Literary World.*

Modern Horsemanship. A New Method of Teaching
Riding and Training by means of pictures from the life. By E. L. ANDERSON. New and Revised Edition containing some observations upon the mode of changing lead in the Gallop. Illustrated by 28 Instantaneous Photographs. Demy 8vo. 21s.

Vice in the Horse and other Papers on Horses and
Riding. By E. L. ANDERSON, Author of "Modern Horsemanship." 1 vol. Demy 8vo, 6s.

The Gallop.
By E. L. ANDERSON. Illustrated by Instantaneous Photography. Fcap. 4to, 2s. 6d.

Scotland in Early Christian Times.

By JOSEPH ANDERSON, LL.D., Keeper of the National Museum of the Antiquaries of Scotland. (Being the Rhind Lectures in Archæology for 1879 and 1880.) 2 vols. Demy 8vo, profusely Illustrated. 12s. each volume.

Contents of Vol. I.—Celtic Churches—Monasteries—Hermitages—Round Towers—Illuminated Manuscripts—Bells—Crosiers—Reliquaries, etc.

Contents of Vol. II.—Celtic Medal-Work and Sculptured Monuments, their Art and Symbolism—Inscribed Monuments in Runics and Oghams—Bilingual Inscriptions, etc.

Scotland in Pagan Times.

By JOSEPH ANDERSON, LL.D. (Being the Rhind Lectures in Archæology for 1881 and 1882.) In 2 vols. Demy 8vo, profusely Illustrated. 12s. each volume.

Contents of Vol. I.—THE IRON AGE.—Viking Burials and Hoards of Silver and Ornaments—Arms, Dress, etc., of the Viking Time—Celtic Art of the Pagan Period—Decorated Mirrors—Enamelled Armlets—Architecture and Contents of the Brochs—Lake-Dwellings—Earth Houses, etc.

Contents of Vol. II.—THE BRONZE AND STONE AGES.—Cairn Burial of the Bronze Age and Cremation Cemeteries—Urns of Bronze Age Types—Stone Circles—Stone Settings—Gold Ornaments—Implements and Weapons of Bronze—Cairn Burial of the Stone Age—Chambered Cairns—Urns of Stone-Age Types—Impliments and Weapons of Stone.

Crofts and Farms in the Hebrides:

Being an account of the Management of an Island Estate for 130 Years. By the DUKE OF ARGYLL. Demy 8vo, 88 pages, 1s.

Continuity and Catastrophes in Geology.

An Address to the Edinburgh Geological Society on its Fiftieth Anniversary, 1st November 1883. By the DUKE OF ARGYLL. Demy 8vo, 1s.

The History of Liddesdale, Eskdale, Ewesdale, Wauchopedale, and the Debateable Land.

Part I. from the Twelfth Century to 1530. By ROBERT BRUCE ARMSTRONG. The edition is limited to 275 copies demy quarto, and 105 copies on large paper (10 inches by 13). With an Appendix of 70 Documents, arranged in Chronological order down to 1566. The selection has been made from private Charter-chests, MS. collections in London and Edinburgh, and rare printed works, and comprises Charters, Rent-rolls, Excerpts from the Accounts of the Lord High Treasurer, Bonds of Manrent, Bonds for the Re-entry of Prisoners, Lists of Scottish Borderers under English Assurance, Injuries inflicted by the English and by the Scottish Borderers under English Assurance from September 1543 to June 1544, interesting Letters and a Military Report on the West March of Scotland and Liddesdale by an English official, etc., etc.

The Volume is illustrated by Maps, Etchings, Lithographs, and Woodcuts, all of which—with the exception of Blaeu's Maps of Liddesdale and Eskdale, and the etchings of James IV., James V., and the Earl of Angus, by C. Lawrie—are either from the author's drawings or wholly executed by himself. The lithographs in colour include facsimiles of four interesting representations of Scottish Border Castles and Towns drawn between the years 1563 and 1566, Plates of Arms of the Lords of Liddesdale, of the Clans of the District, of Lindsay of Wauchope, also of the Seals of John Armstrong and William Elliot, etc., etc. 42s. and 84s.

Morning Clouds:

Being divers Poems by H. B. BAILDON, B.A. Cantab., Author of "Rosamund," etc. Ex. fcap. 8vo, 5s.

By the same Author.

First Fruits. 5s.   Rosamund. 5s.

Dr. Heidenhoff's Process.

By EDWARD BELLAMY. Crown 8vo, 6s.

Miss Ludington's Sister: a Romance of Immortality.

By EDWARD BELLAMY, Author of "Dr. Heidenhoff's Process." Crown 8vo, 6s.

LIST OF BOOKS

The Voyage of the Paper Canoe.
A Geographical Journey of 2500 miles, from Quebec to the Gulf of Mexico, during the year 1874-75. By N. H. BISHOP. With Maps and Plates. Demy 8vo, 10s. 6d.

On Self-Culture:
Intellectual, Physical, and Moral. A *Vade-Mecum* for Young Men and Students. By JOHN STUART BLACKIE, Emeritus Professor of Greek in the University of Edinburgh. Fifteenth Edition. Fcap. 8vo, 2s. 6d.

"Every parent should put it into the hands of his son."—*Scotsman*.

"Students in all countries would do well to take as their *vade-mecum* a little book on self-culture by the eminent Professor of Greek in the University of Edinburgh."—*Medical Press and Circular.*

"An invaluable manual to be put into the hands of students and young men."—*Era*.

"Written in that lucid and nervous prose of which he is a master."—*Spectator*.

"An adequate guide to a generous, eager, and sensible life."—*Academy*.

"The volume is a little thing, but it is a *multum in parvo* . . . a little locket gemmed within and without with real stones fitly set."—*Courant*.

By the same Author.

On Greek Pronunciation. Demy 8vo, 3s. 6d.
On Beauty. Crown 8vo, cloth, 8s. 6d.
Lyrical Poems. Crown 8vo, cloth, 7s. 6d.
The Language and Literature of the Scottish Highlands. Crown 8vo, 6s.
Four Phases of Morals:
Socrates, Aristotle, Christianity, and Utilitarianism. Lectures delivered before the Royal Institution, London. Ex. fcap. 8vo, Second Edition, 5s.

Songs of Religion and Life. Fcap. 8vo, 6s.
Musa Burschicosa.
A book of Songs for Students and University Men. Fcap. 8vo, 2s. 6d.

War Songs of the Germans. Fcap. 8vo, 2s. 6d. cloth; 2s. paper.
Political Tracts. No. 1. GOVERNMENT. No. 2. EDUCATION. 1s. each.
Gaelic Societies. Highland Depopulation and Land Law Reform. Demy 8vo, 6d.
Homer and the Iliad.
In three Parts. 4 vols. Demy 8vo, 42s.

Love Revealed: Meditations on the Parting Words of
Jesus with His Disciples, in John xiii-xvii. By the Rev. GEORGE BOWEN, Missionary at Bombay. Small 4to, 5s.

"No true Christian could put the book down without finding in himself more traces of the blessed unction which drops from every page."—*Record*.

"Here is a feast of fat things, of fat things full of marrow."—*Sword and Trowel*.

"A more stimulating work of its class has not appeared for many a long day."—*Scotsman*.

"The present work is eminently qualified to help the devotional life."—*Literary World*.

"He writes plainly and earnestly, and with a true appreciation of the tender beauties of what are really among the finest passages in the New Testament."—*Glasgow Herald*.

"Verily, Verily," The Amens of Christ.
By the Rev. GEORGE BOWEN, Missionary at Bombay. Small 4to, cloth, 5s.

"For private and devotional reading this book will be found very helpful and stimulative."—*Literary World*.

PUBLISHED BY DAVID DOUGLAS. 5

Daily Meditations by Rev. George Bowen, Missionary at Bombay. With Introductory Notice by Rev. W. HANNA, D.D., Author of "The Last Day of our Lord's Passion." New Edition. Small 4to, cloth, 5s.

"These meditations are the production of a missionary whose mental history is very remarkable. . . . His conversion to a religious life is undoubtedly one of the most remarkable on record. They are all distinguished by a tone of true piety, and are wholly free from a sectarian or controversial bias."—*Morning Post.*

Works by John Brown, M.D., F.R.S.E.
HORÆ SUBSECIVÆ. 3 Vols.
I. Locke and Sydenham. Fifth Edition, with Portrait by James Faed. Crown 8vo, 7s. 6d.
II. Rab and his Friends. Thirteenth Edition. Crown 8vo, 7s. 6d.
III. John Leech. Fifth Edition, with Portrait by George Reid, R.S.A. Crown 8vo, 7s. 6d.

Separate Papers, extracted from "Horæ Subsecivæ."
RAB AND HIS FRIENDS. With India-proof Portrait of the Author after Faed, and seven India-proof Illustrations after Sir G. Harvey, Sir J. Noel Paton, Mrs. Blackburn, and G. Reid, R.S.A. Demy 4to, cloth, 9s.
MARJORIE FLEMING: A Sketch. Being a Paper entitled "Pet Majorie: A Story of a Child's Life fifty years ago." New Edition, with Illustrations. Demy 4to, 7s. 6d. and 6s.
RAB AND HIS FRIENDS. Cheap Illustrated Edition. Square 12mo, ornamental wrapper, 1s.
LETTER TO THE REV. JOHN CAIRNS, D.D. Second Edition, crown 8vo, sewed, 2s.
ARTHUR H. HALLAM. Fcap., sewed, 2s.; cloth, 2s. 6d.
RAB AND HIS FRIENDS. Sixty-sixth thousand. Fcap., sewed, 6d.
MARJORIE FLEMING: A Sketch. Sixteenth Thousand. Fcap., sewed, 6d.
OUR DOGS. Twentieth thousand. Fcap., sewed, 6d.
"WITH BRAINS, SIR." Seventh thousand. Fcap., sewed, 6d.
MINCHMOOR. Tenth Thousand. Fcap., sewed, 6d.
JEEMS THE DOOR-KEEPER: A Lay Sermon. Twelfth thousand. Price 6d.
THE ENTERKIN. Seventh Thousand. Price 6d.
PLAIN WORDS ON HEALTH. Twenty-seventh thousand. Price 6d.
SOMETHING ABOUT A WELL: WITH MORE OF OUR DOGS. Price 6d.

From Schola to Cathedral. A Study of Early Christian Architecture in its relation to the life of the Church. By G. BALDWIN-BROWN, Professor of Fine Arts in the University of Edinburgh. Demy 8vo, Illustrated, 7s. 6d.

The book treats of the beginnings of Christian Architecture, from the point of view of recent discoveries and theories, with a special reference to the outward resemblance of early Christian communities to other religious associations of the time.

The Capercaillie in Scotland.
By J. A. HARVIE-BROWN. Etchings on Copper, and Map Illustrating the extension of its range since its Restoration at Taymouth in 1837 and 1838. Demy 8vo, 8s. 6d.

The History of Selkirkshire.
By T. CRAIG-BROWN. Two Vols. Demy 4to. [*In the Press.*

Pugin Studentship Drawings. Being a selection from Sketches, Measured Drawings, and details of Domestic and Ecclesiastic Buildings in England and Scotland. By G. WASHINGTON-BROWNE, F.S.A. Scot., Architect. 1 vol. Folio, Illustrated. [*In the Press.*

Select Hymns for Church and Home.
By R. BROWN-BORTHWICK. Second Edition. 16mo, 1s. 6d.

"The Red Book of Menteith" Reviewed.
By GEORGE BURNETT, Advocate, Lyon King of Arms. Small 4to, 5s.

John Burroughs's Essays.

Six Books of Nature, Animal Life, and Literature. Choice Edition. Revised by the Author. 6 vols. 32mo, cloth, 12s.; or in smooth ornamental wrappers, 6s.; or separately at 1s. each vol., or 2s. in cloth.

WINTER SUNSHINE.	FRESH FIELDS.
LOCUSTS AND WILD HONEY.	BIRDS AND POETS.
WAKE-ROBIN.	PEPACTON.

"Whichever essay I read, I am glad I read it, for pleasanter reading, to those who love the country, with all its enchanting sights and sounds, cannot be imagined."—*Spectator.*

"Mr. Burroughs is one of the most delightful of American Essayists, steeped in culture to the finger ends."—*Pall Mall Gazette.*

FRESH FIELDS. By JOHN BURROUGHS. Library Edition. Crown 8vo, 6s.

Dr. Sevier: A Novel.

By GEO. W. CABLE, Author of "Old Creole Days," etc. In 2 vols., crown 8vo, price 12s.

Old Creole Days.

By GEO. W. CABLE. 32mo, 1s.; and in Cloth, 2s.

"We cannot recall any contemporary American writer of fiction who possesses some of the best gifts of the novelist in a higher degree."—*St. James's Gazette.*

Memoir of John Brown, D.D.

By JOHN CAIRNS, D.D., Berwick-on-Tweed. Crown 8vo, 7s. 6d.

My Indian Journal.

Containing Descriptions of the principal Field Sports of India, with Notes on the Natural History and Habits of the Wild Animals of the Country. By Colonel WALTER CAMPBELL, Author of "The Old Forest Ranger." Small demy 8vo, with Illustrations by Wolf, 16s.

Life and Works of Rev. Thomas Chalmers, D.D., LL.D.

MEMOIRS OF THE REV. THOMAS CHALMERS. By Rev. W. HANNA, D.D., LL.D. New Edition. 2 vols. crown 8vo, cloth, 12s.

DAILY SCRIPTURE READINGS. Cheap Edition. 2 vols. crown 8vo, 10s.

ASTRONOMICAL DISCOURSES, 1s.

COMMERCIAL DISCOURSES, 1s.

SELECT WORKS, in 12 vols., crown 8vo, cloth, per vol. 6s.
- Lectures on the Romans. 2 vols.
- Sermons. 2 vols.
- Natural Theology, Lectures on Butler's Analogy, etc. 1 vol.
- Christian Evidences, Lectures on Paley's Evidences, etc. 1 vol.
- Institutes of Theology. 2 vols.
- Political Economy, with Cognate Essays. 1 vol.
- Polity of a Nation. 1 vol.
- Church and College Establishments. 1 vol.
- Moral Philosophy, Introductory Essays, Index, etc. 1 vol.

Lectures on Surgical Anatomy.

By JOHN CHIENE, M.D., Professor of Surgery in the University of Edinburgh. In demy 8vo. With numerous Illustrations drawn on Stone by BERJEAU. 12s. 6d.

"The book will be a great help to both teachers and taught, and students can depend upon the teaching as being sound."—*Medical Times and Gazette.*

Lectures on the Elements or First Principles of Surgery.

By JOHN CHIENE, M.D., Professor of Surgery in the University of Edinburgh. Demy 8vo, 2s. 6d.

Traditional Ballad Airs.

Arranged and Harmonised for the Pianoforte and Harmonium. By W. CHRISTIE, M.A., and the late WILLIAM CHRISTIE, Monquhitter. Demy 4to, Vols. I. and II. 42s. each.

PUBLISHED BY DAVID DOUGLAS. 7

Archibald Constable and his Literary Correspondents:
a Memorial. By his Son, THOMAS CONSTABLE. 3 vols. demy 8vo, 36s., with Portrait.

"He (Mr. Constable) was a genius in the publishing world. . . . The creator of the Scottish publishing trade."—*Times.*

The Dandie Dinmont Terrier: Its History and Characteristics. Compiled from the most Authentic Sources. By CHARLES COOK. Illustrated by Portraits of Authentic Specimens of the Pure Breed. Drawn and Etched by W. HOLE, A.R.S.A.

"His history of various celebrated Dandies, his pedigrees, and his anecdotes of Dandies' doings, are so well compiled and so interesting, that every one who owns a pet Dandie, let alone all who take an interest in dogs in general, will read the book with great interest."—*Pall Mall Gazette.*

"Mr. Cook's work teems with interest, and, quite unique of its kind, it will no doubt be found on the drawing-room table of all who take an interest in this particular strain of Scottish terriers."—*Field.*

The Earldom of Mar, in Sunshine and in Shade, during Five Hundred Years. With incidental Notices of the leading Cases of Scottish Dignities of King Charles I. till now. By ALEXANDER, EARL OF CRAWFORD AND BALCARRES, LORD LINDSAY, etc. etc. 2 vols. demy 8vo, 32s.

The Crime of Henry Vane: a Study with a Moral.
By J. S. of Dale, Author of "Guerndale." Crown 8vo, 6s.

A Clinical and Experimental Study of the Bladder during Parturition. By J. H. CROOM, M.B., F.R.C.P.E. Small 4to, with Illustrations, 6s.

Wild Men and Wild Beasts.
Adventures in Camp and Jungle. By Lieut.-Colonel GORDON CUMMING. With Illustrations by Lieut.-Col. BAIGRIE and others. Small 4to, 24s.
Also a cheaper edition, with *Lithographic* Illustrations. 8vo, 12s.

Prue and I.
By GEORGE WILLIAM CURTIS. 32mo, 1s. paper; or 2s. cloth extra.

Contents.—Dinner Time—My Chateaux—Sea from Shore—Titbottom's Spectacles—A Cruise in the Flying Dutchman—Family Portraits—Our Cousin the Curate.

"This is a dainty piece of work, and well deserved reprinting."—*Athenæum.*
"These charming sketches will be enjoyed by all cultured readers."—*Daily Chronicle.*

Burnt Njal.
From the Icelandic of the Njal's Saga. By Sir GEORGE WEBBE DASENT, D.C.L. 2 vols. demy 8vo, with Maps and Plans, 28s.

Gisli the Outlaw.
From the Icelandic. By Sir GEORGE WEBBE DASENT, D.C.L. Small 4to, with Illustrations, 7s. 6d.

A Daughter of the Philistines: A Novel.
Crown 8vo, 6s. Also a cheaper edition in paper binding, 2s.

"The story is very powerfully told, possesses a piquantly satirical flavour, and possesses the very real attraction of freshness."—*Scotsman.*
"It is cleverly and brightly written."—*Academy.*

A Manual of Chemical Analysis.
By Professor WILLIAM DITTMAR. Ex. fcap. 8vo, 5s.

TABLES FORMING AN APPENDIX TO DITTO. Demy 8vo, 3s. 6d.

A Chat in the Saddle; or Patroclus and Penelope.
By THEO. A. DODGE, Lieut.-Colonel, United States Army. Illustrated by 14 Instantaneous Photographs. Demy 8vo, half-leather binding, 21s.

Veterinary Medicines; their Actions and Uses.
By FINLAY DUN. Sixth Edition, revised and enlarged. Demy 8vo, 15s.

Social Life in Former Days;
Chiefly in the Province of Moray. Illustrated by Letters and Family Papers. By E. DUNBAR DUNBAR, late Captain 21st Fusiliers. 2 vols. Demy 8vo, 19s. 6d.

Letters of Thomas Erskine of Linlathen.
Edited by WILLIAM HANNA, D.D., Author of the "Memoirs of Dr. Chalmers," etc. Fourth Edition. Crown 8vo, 7s. 6d.

The Unconditional Freeness of the Gospel.
By THOMAS ERSKINE of Linlathen. New Edition, revised. Crown 8vo, 3s. 6d.

By the same Author.

The Brazen Serpent:
Or, Life coming through Death. Third Edition. Crown 8vo, 5s.

The Internal Evidence of Revealed Religion.
Crown 8vo, 5s.

The Spiritual Order,
And other Papers selected from the MSS. of the late THOMAS ERSKINE of Linlathen. Third Edition. Crown 8vo, 5s.

The Doctrine of Election,
And its Connection with the General Tenor of Christianity, illustrated especially from the Epistle to the Romans. Second Edition. Crown 8vo, 6s.

Three Visits to America.
By EMILY FAITHFULL. Demy 8vo, 9s.

Twelve Sketches of Scenery and Antiquities on the
Line of the Great North of Scotland Railway. By GEORGE REID, R.S.A. With Illustrative Letterpress by W. FERGUSON of Kinmundy. 4to, 15s.

Guide to the Great North of Scotland Railway.
By W. FERGUSON of Kinmundy. Crown 8vo; in paper cover, 1s.; cloth cover, 1s. 6d.

Letters and Journals of Mrs. Calderwood of Polton,
from England, Holland, and the Low Countries, in 1756. Edited by ALEX. FERGUSSON, Lieut.-Colonel, Author of "Henry Erskine and his Kinsfolk." Demy 8vo, Illustrated, 18s.

The Laird of Lag; A Life-Sketch of Sir Robert Grierson.
By ALEX. FERGUSSON, Lieut.-Colonel, Author of "Henry Erskine and his Kinsfolk," "Mrs. Calderwood's Journey." Demy 8vo, with Illustrations, 12s.

Autobiography of Mrs. Fletcher
(of Edinburgh), with Letters and other Family Memorials. Edited by her Daughter. Third Edition. Crown 8vo, 7s. 6d.

L'Histoire de France.
Par M. LAME FLEURY. New Edition, corrected to 1883. 18mo, cloth, 2s. 6d.

The Deepening of the Spiritual Life.
By A. P. FORBES, D.C.L., Bishop of Brechin. Seventh Edition. Paper, 1s.; cloth, 1s. 6d. Calf, red edges, 3s. 6d.

Kalendars of Scottish Saints,

With Personal Notices of those of Alba, etc. By ALEXANDER PENROSE FORBES, D.C.L., Bishop of Brechin. 1 vol. 4to, price £3, 3s. A few copies for sale on large paper, £5, 15s. 6d.

"A truly valuable contribution to the archæology of Scotland."—*Guardian.*

"We must not forget to thank the author for the great amount of information he has put together, and for the labour he has bestowed on a work which can never be remunerative."—*Saturday Review.*

"His laborious and very interesting work on the early Saints of Alba, Laudonia, and Strathclyde."—*Quarterly Review.*

Missale Drummondiense. The Ancient Irish Missal in

the possession of the Baroness Willoughby d'Eresby. Edited by the Rev. G. H. FORBES. Half-Morocco, Demy 8vo, 12s.

Forestry and Forest Products.

Prize Essays of the Edinburgh International Forestry Exhibition 1884. Edited by JOHN RATTRAY M.A., B.Sc., and HUGH ROBERT MILL, B.Sc. Demy 8vo, Illustrated, 9s.

Fragments of Truth:

Being the Exposition of several Passages of Scripture. Third Edition. Ex. fcap. 8vo, 5s.

Studies in English History.

By JAMES GAIRDNER and JAMES SPEDDING. Demy 8vo, 12s.

Contents.—The Lollards—Sir John Falstaff—Katherine of Arragon's First and Second Marriages—Case of Sir Thomas Overbury—Divine Right of Kings—Sunday, Ancient and Modern.

Gifts for Men.

By X. H. Crown 8vo, 6s.

"There is hardly a living theologian who might not be proud to claim many of her thoughts as his own."—*Glasgow Herald.*

Sketches Literary and Theological:

Being Selections from the unpublished MSS. of the Rev. GEORGE GILFILLAN. Edited by FRANK HENDERSON, Esq., M.P. Demy 8vo, 7s. 6d.

The Roof of the World:

Being the Narrative of a Journey over the High Plateau of Tibet to the Russian Frontier, and the Oxus Sources on Pamir. By Brigadier-General T. E. GORDON, C.S.I. With numerous Illustrations. Royal 8vo, 31s. 6d.

Works by Margaret Maria Gordon (née Brewster).

THE HOME LIFE OF SIR DAVID BREWSTER. By his Daughter. Second Edition. Crown 8vo, 6s. Also a cheaper Edition. Crown 8vo, 2s. 6d.

WORK; Or, Plenty to do and How to do it. Thirty-Sixth Thousand. Fcap. 8vo, cloth, 2s. 6d.

"A better gift-book for young domestic servants we do not know."—*Literary Gazette.*

WORKERS. Fourth Thousand. Fcap. 8vo, limp cloth, 1s.

LITTLE MILLIE AND HER FOUR PLACES. Cheap Edition. Fifty-eighth Thousand. Limp cloth, 1s.

SUNBEAMS IN THE COTTAGE; Or, What Women may Do. A Narrative chiefly addressed to the Working Classes. Cheap Edition. Forty-fifth Thousand. Limp cloth, 1s.

PREVENTION; Or, An Appeal to Economy and Common Sense. 8vo, 6d.

THE WORD AND THE WORLD. Twelfth Edition. 2d.

LEAVES OF HEALING FOR THE SICK AND SORROWFUL. Cheap Edition, limp cloth, 2s.

THE MOTHERLESS BOY. With an Illustration by Sir NOEL PATON, R.S.A. Cheap Edition, limp cloth, 1s.

LIST OF BOOKS

OUR DAUGHTERS; An Account of the Young Women's Christian Association and Institute Union. 2d.

HAY MACDOWALL GRANT OF ARNDILLY; His Life, Labours, and Teaching. New and Cheaper Edition. 1 vol. crown 8vo, limp cloth, 2s. 6d.

Ladies' Old-Fashioned Shoes.
By T. WATSON GREIG, of Glencarse. Folio, illustrated by 11 Chromolithographs. 31s. 6d.

The Life of our Lord.
By the Rev. WILLIAM HANNA, D.D., LL.D. 6 vols., handsomely bound in cloth extra, gilt edges, 30s.
Separate vols., cloth extra, gilt edges, 5s. each.
1. THE EARLIER YEARS OF OUR LORD. Fifth Edition.
2. THE MINISTRY IN GALILEE. Fourth Edition.
3. THE CLOSE OF THE MINISTRY. Sixth Thousand.
4. THE PASSION WEEK. Sixth Thousand.
5. THE LAST DAY OF OUR LORD'S PASSION. Twenty-third Edition.
6. THE FORTY DAYS AFTER THE RESURRECTION. Eighth Edition.

The Resurrection of the Dead.
By WILLIAM HANNA, D.D., LL.D. Second Edition. Fcap. 8vo, 5s.

Mingo, and other Sketches in Black and White.
By JOEL CHANDLER HARRIS (*Uncle Remus*). 32mo, 1s.; and in cloth, 2s.

Notes of Caithness Family History.
By the late JOHN HENDERSON, W.S. 4to, 31s. 6d.

Errors in the Use of English.
Illustrated from the Writings of English Authors, from the Fourteenth Century to Our own Time. By the late W. B. HODGSON, LL.D., Professor of Political Economy in the University of Edinburgh. Fifth Edition. Crown 8vo, 3s. 6d.

"Those who most need such a book as Dr. Hodgson's will probably be the last to look into it. It will certainly amuse its readers, and will probably teach them a good deal which they did not know, or at least never thought about, before."—*Saturday Review.*

"His conversation, as every one who had the pleasure of his acquaintance knows, sparkled with anecdote and epigram, and not a little of the lustre and charm of his talk shines out of those pages."—*The Scotsman.*

Life and Letters of W. B. Hodgson, LL.D., late Professor of Political Economy in the University of Edinburgh. Edited by Professor J. M. D. MEIKLEJOHN, M.A. Crown 8vo, 7s. 6d.

Sketches: Personal and Pensive.
By WILLIAM HODGSON. Fcap. 8vo, 2s. 6d.

"Quasi Cursores." Portraits of the High Officers and Professors of the University of Edinburgh. Drawn and Etched by WILLIAM HOLE, A.R.S.A. The book is printed on beautiful hand-made paper by Messrs. T. & A. Constable. It contains 45 Plates (64 Portraits), with Biographical Notices of all the present Incumbents. The impression is strictly limited. Quarto Edition (750 Copies only for sale), £2, 10s. Folio Edition, Japan Proofs (100 Copies only for sale), £5, 10s.

"A work of great value and of high artistic merit, not merely in respect of the portraits, but also in respect of the typography, the paper, the binding, and the general get-up of the volume. ... It does great credit to the resources of Edinburgh, both as a seat of learning, and as a centre of literary production on its mechanical side."—*Times.*

PUBLISHED BY DAVID DOUGLAS. 11

The Breakfast Table Series.
In 6 vols. By OLIVER WENDELL HOLMES. New and Revised Editions, containing Prefaces and additional Bibliographical Notes by the Author.
Every man his own Boswell.
THE AUTOCRAT OF THE BREAKFAST TABLE. 2 vols., 2s.
THE POET AT THE BREAKFAST TABLE. 2 vols., 2s.
THE PROFESSOR AT THE BREAKFAST TABLE. 2 vols, 2s.
   Also bound in dark blue cloth, gilt top, at 2s. a volume, or in a neat cloth box, at 15s.
   "Small enough to be carried in any sensibly constructed pocket, clear enough in type to accommodate any fastidious eyesight, pleasant and instructive enough for its perusal to be undertaken with the certainty of present enjoyment and the prospect of future profit."—*Whitehall Review.*
   Also a LIBRARY EDITION, in 3 vols. Crown 8vo, printed at the Riverside Press, Cambridge, with a Steel Portrait of the Author, 10s. 6d. each volume.

Traces in Scotland of Ancient Water Lines, Marine,
Lacustrine, and Fluviatile. By DAVID MILNE-HOME, LL.D., F.R.S.E. Demy 8vo, 3s. 6d.

A Sketch of the Life of George Hope of Fenton Barns.
Compiled by his DAUGHTER. Crown 8vo, 6s.

One Summer.
By BLANCHE WILLIS HOWARD. 32mo, paper, 1s.; cloth, 1s. 6d. and 2s.

W. D. Howells' Writings in "American Author" Series.
THE RISE OF SILAS LAPHAM. 2 vols. 16mo, 2s.
A FOREGONE CONCLUSION. 1 vol., 1s.
A CHANCE ACQUAINTANCE. 1 vol., 1s.
THEIR WEDDING JOURNEY. 1 vol., 1s.
A COUNTERFEIT PRESENTMENT, AND THE PARLOUR CAR. 1 vol., 1s.
THE LADY OF THE AROOSTOOK. 2 vols., 2s.
OUT OF THE QUESTION, and AT THE SIGN OF THE SAVAGE. 1 vol. 1s.
THE UNDISCOVERED COUNTRY. 2 vols., 2s.
A FEARFUL RESPONSIBILITY, and TONELLI'S MARRIAGE. 1 vol., 1s.
VENETIAN LIFE. 2 vols., 2s.
ITALIAN JOURNEYS. 2 vols., 2s.
   All the above may be had in cloth at 2s. each vol.
         Copyright Library Edition.
A MODERN INSTANCE. 2 vols., 12s.
A WOMAN'S REASON. 2 vols., 12s.
DR. BREEN'S PRACTICE. 1 vol., 3s. 6d.
INDIAN SUMMER. 1 vol., 6s.

A Memorial Sketch, and a Selection from the Letters
of the late Lieut. JOHN IRVING, R.N., of H.M.S. "Terror," in Sir John Franklin's Expedition to the Arctic Regions. Edited by BENJAMIN BELL, F.R.C.S.E. With Facsimiles of the Record and Irving's Medal and Map. Post 8vo, 5s.

Jack and Mrs. Brown, and other Stories.
By the Author of "Blindpits." Crown 8vo, paper, 2s. 6d.; cloth, 3s. 6d.

Zeph: A Posthumous Story.
By HELEN JACKSON ("H.H."). Author of "Ramona," etc. Crown 8vo, 6s.

Epitaphs and Inscriptions from Burial-Grounds and
Old Buildings in the North-East of Scotland. By the late ANDREW JERVISE, F.S.A. Scot. With a Memoir of the Author. Vol. II. Cloth, small 4to, 32s.
   Do.   do.   Roxburghe Edition, 42s.

LIST OF BOOKS

The History and Traditions of the Land of the Lindsays
in Angus and Mearns. New Edition, Edited and Revised by the Rev. JAMES GAMMACK, M.A. Demy 8vo, 14s.
Do. do. Large Paper Edition (of which only 50 are printed), Demy 4to, Roxburghe binding, 42s.

Memorials of Angus and the Mearns: an Account,
Historical, Antiquarian, and Traditionary, of the Castles and Towns visited by Edward I., and of the Barons, Clergy, and others who swore Fealty to England in 1291-6. By the late ANDREW JERVISE, F.S.A. Scot. Rewritten and corrected by the Rev. JAMES GAMMACK, M.A. Illustrated with Etchings by W. HOLE, A.R.S.A. 2 vols. Demy 8vo, 28s.; Large Paper, 2 vols. Demy 4to, 63s.

Pilate's Question, "Whence art Thou?"
An Essay on the Personal Claims asserted by Jesus Christ, and how to account for them. By JOHN KENNEDY, M.A., D.D., London. Crown 8vo, 3s. 6d.

Sermons by the Rev. John Ker, D.D., Glasgow.
Thirteenth Edition. Crown 8vo, 6s.
"A very remarkable volume of sermons."—*Contemporary Review.*
"The sermons before us are of no common order; among a host of competitors they occupy a high class—we were about to say the highest class—whether viewed in point of composition, or thought, or treatment."—*B. and F. Evangelical Review.*

The English Lake District as interpreted in the Poems
of Wordsworth. By WILLIAM KNIGHT, Professor of Moral Philosophy in the University of St. Andrews. Ex. fcap. 8vo, 5s.

Colloquia Peripatetica (Deep Sea Soundings):
Being Notes of Conversations with the late John Duncan, LL.D., Professor of Hebrew in the New College, Edinburgh. By WILLIAM KNIGHT, Professor of Moral Philosophy in the University of St. Andrews. Fifth Edition, enlarged, 5s.

Lindores Abbey, and the Burgh of Newburgh;
Their History and Annals. By ALEXANDER LAING, LL.D., F.S.A. Scot. Small 4to. With Index, and thirteen Full-page and ten Woodcut Illustrations, 21s.
"This is a charming volume in every respect."—*Notes and Queries.*
"The prominent characteristics of the work are its exhaustiveness and the thoroughly philosophic spirit in which it is written."—*Scotsman.*

Recollections of Curious Characters and Pleasant Places. By CHARLES LANMAN, Washington; Author of "Adventures in the Wilds of America," "A Canoe Voyage up the Mississippi," "A Tour to the River Saguenay," etc. etc. Small Demy 8vo, 12s.

Essays and Reviews.
By the late HENRY H. LANCASTER, Advocate; with a Prefatory Notice by the Rev. B. JOWETT, Master of Balliol College, Oxford. Demy 8vo, with Portrait, 14s.

An Echo of Passion.
By GEO. PARSONS LATHROP. 32mo, 1s.; and in cloth, 2s.

On the Philosophy of Ethics. An Analytical Essay.
By S. S. LAURIE, A.M., F.R.S.E., Professor of the Theory, History, and Practice of Education in the University of Edinburgh. Demy 8vo, 6s.

Notes on British Theories of Morals.
By Prof. S. S. LAURIE. Demy 8vo, 6s.

Leaves from the Buik of the West Kirke.
By GEO. LORIMER. With a Preface by the Rev. JAS. MACGREGOR, D.D. 4to.

Bible Studies in Life and Truth.
By the Rev. ROBERT LORIMER, M.A., Free Church, Mains and Strathmartine. Crown 8vo, 5s.

Sermons.
By Rev. ADAM LIND, M.A., Elgin. Ex. fcap. 8vo, 5s.

Only an Incident.
A Novel. By Miss G. D. LITCHFIELD. Crown 8vo, 6s.

A Lost Battle.
A Novel. 2 vols. Crown 8vo, 17s.
"This in every way remarkable novel."—*Morning Post.*
"We are all the more ready to do justice to the excellence of the author's drawing of characters."—*Athenæum.*

John Calvin, a Fragment by the Late Thomas M'Crie,
Author of "The Life of John Knox." Demy 8vo, 6s.

The Parish of Taxwood, and some of its Older Memories.
By Rev. J. R. MACDUFF, D.D. Extra fcap. 8vo, illustrated, 3s. 6d.

Principles of the Algebra of Logic, with Examples.
By ALEX. MACFARLANE, M.A., D.Sc. (Edin.), F.R.S.E. 5s.

Memoir of Sir James Dalrymple, First Viscount Stair.
A Study in the History of Scotland and Scotch Law during the Seventeenth Century. By Æ. J. G. MACKAY, Advocate. 8vo, 12s.

Storms and Sunshine of a Soldier's Life.
Lt.-General COLIN MACKENZIE, C.B., 1825-1881. With a Portrait. 2 vols. Crown 8vo, 15s.
"A very readable biography . . . of one of the bravest and ablest officers of the East India Company's army."—*Saturday Review.*

Nugæ Canoræ Medicæ.
Lays of the Poet Laureate of the New Town Dispensary. Edited by Professor DOUGLAS MACLAGAN. 4to, with Illustrations, 7s. 6d.

The Hill Forts, Stone Circles, and other Structural Remains of Ancient Scotland. By C. MACLAGAN, Lady Associate of the Society of Antiquaries of Scotland. With Plans and Illustrations. Folio, 31s. 6d.
"We need not enlarge on the few inconsequential speculations which rigid archæologists may find in the present volume. We desire rather to commend it to their careful study, fully assured that not only they, but also the general reader, will be edified by its perusal."—*Scotsman.*

The Light of the World.
By DAVID M'LAREN, Minister of Humbie. Crown 8vo, 6s.

The Book of Psalms in Metre.
According to the version approved of by the Church of Scotland. Revised by Rev. DAVID M'LAREN. Crown 8vo, 7s. 6d.

Omnipotence belongs only to the Beloved.
By Mrs. BREWSTER MACPHERSON. Extra fcap., 3s. 6d.

In Partnership. Studies in Story-Telling
By BRANDER MATTHEWS and H. C. BUNNER. 32mo, 1s. in paper, and 2s. in cloth.

Antwerp Delivered in MDLXXVII.:
A Passage from the History of the Netherlands, illustrated with Facsimiles of a rare series of Designs by Martin de Vos, and of Prints by Hogenberg, the Lierixes, etc. By Sir WILLIAM STIRLING-MAXWELL, Bart., K.T. and M.P. In 1 vol. Folio, 5 guineas.
"A splendid folio in richly ornamented binding, protected by an almost equally ornamental slip-cover. . . . Remarkable illustrations of the manner in which the artists of the time 'pursued their labours in a country ravaged by war, and in cities ever menaced by siege and sack.'"—*Scotsman.*

The History of Old Dundee, narrated out of the Town
Council Register, with Additions from Contemporary Annals. By ALEXANDER MAXWELL, F.S.A. Scot. 4to. (To subscribers, 21s.)

Researches and Excavations at Carnac (Morbihan),
The Bossenno, and Mont St. Michel. By JAMES MILN. Royal 8vo, with Maps, Plans, and numerous Illustrations in Wood-Engraving and Chromolithography.

Excavations at Carnac (Brittany), a Record of Archæ-ological Researches in the Alignments of Kermario. By JAMES MILN. Royal 8vo, with Maps, Plans, and numerous Illustrations in Wood-Engraving. 15s.

The Past in the Present—What is Civilisation?
Being the Rhind Lectures in Archæology, delivered in 1876 and 1878. By ARTHUR MITCHELL, M.D., LL.D., Secretary to the Society of Antiquaries of Scotland. In 1 vol. Demy 8vo, with 148 Woodcuts, 15s.

"Whatever differences of opinion, however, may be held on minor points, there can be no question that Dr. Mitchell's work is one of the ablest and most original pieces of archæological literature which has appeared of late years."—*St. James's Gazette.*

In War Time. A Novel.
By S. WEIR MITCHELL. Crown 8vo, 6s.

Our Scotch Banks:
Their Position and their Policy. By WM. MITCHELL, S.S.C. Third Edition. 8vo, 5s.

On Horse-Breaking.
By ROBERT MORETON. Second Edition. Fcap. 8vo, 1s.

Ecclesiological Notes on some of the Islands of Scot-land, with other Papers relating to Ecclesiological Remains on the Scottish Mainland and Islands. By THOMAS S. MUIR, Author of "Characteristics of Church Architecture," etc. Demy 8vo, with numerous Illustrations, 21s.

Ancient Scottish Lake-Dwellings or Crannogs, with a Supplementary Chapter on Remains of Lake-Dwellings in England. By ROBERT MUNRO, M.D., F.S.A. Scot. 1 vol. demy 8vo, profusely illustrated, 21s.

"A standard authority on the subject of which it treats."—*Times.*
"... Our readers may be assured that they will find very much to interest and instruct them in the perusal of the work."—*Athenæum.*

"The Lenox of Auld:" An Epistolary Review of "The Lennox, by William Fraser." By MARK NAPIER. With Woodcuts and Plates. 4to, 15s.

Tenants' Gain not Landlords' Loss, and some other Economic Aspects of the Land Question. By JOSEPH SHIELD NICHOLSON, M.A., Professor of Political Economy in the University of Edinburgh. Crown 8vo, 5s.

Camps in the Caribbees: Adventures of a Naturalist in the Lesser Antilles. By FREDERICK OBER. Illustrations, Demy 8vo, 12s.

"Well-written and well-illustrated narrative of camping out among the Caribbees."—*Westminster Review.*
"Varied were his experiences, hairbreadth his escapes, and wonderful his gleanings in the way of securing rare birds."—*The Literary World.*

Cookery for the Sick and a Guide for the Sick-Room.
By C. H. OGG, an Edinburgh Nurse. Fcap. 1s.

The Lord Advocates of Scotland from the close of the Fifteenth Century to the passing of the Reform Bill. By G. W. T. OMOND, Advocate. 2 vols. Demy 8vo, 28s.

PUBLISHED BY DAVID DOUGLAS. 15

An Irish Garland.
By Mrs. S. M. B. Piatt. Crown 8vo, 3s. 6d.

The Children Out of Doors. A Book of Verses
By Two in One House. Crown 8vo, 3s. 6d.

Records of the Coinage of Scotland, from the earliest
period to the Union. Collected by R. W. Cochran-Patrick, M.P. Only two hundred and fifty copies printed. Now ready, in 2 vols. 4to, with 16 Full-page Illustrations, Six Guineas.

"The future Historians of Scotland will be very fortunate if many parts of their materials are so carefully worked up for them and set before them in so complete and taking a form."—*Athenæum.*

"When we say that these two volumes contain more than 770 records, of which more than 550 have never been printed before, and that they are illustrated by a series of Plates, by the autotype process, of the coins themselves, the reader may judge for himself of the learning, as well as the pains, bestowed on them both by the Author and the Publisher."—*Times.*

"The most handsome and complete Work of the kind which has ever been published in this country."—*Numismatic Chronicle, Pt. IV.*, 1876.

Early Records relating to Mining in Scotland:
Collected by R. W. Cochran-Patrick, M.P. Demy 4to, 31s. 6d.

"The documents . . . comprise a great deal that is very curious, and no less that will be important to the historian in treating of the origin of one of the most important branches of the national industry."—*Daily News.*

"Such a book . . . revealing as it does the first developments of an industry which has become the mainspring of the national prosperity, ought to be specially interesting to all patriotic Scotchmen."—*Saturday Review.*

The Medals of Scotland: a Descriptive Catalogue of
the Royal and other Medals relating to Scotland. By R. W. Cochran-Patrick, M.P. Dedicated by special permission to Her Majesty the Queen. Demy 4to, with plates in facsimile of all the principal pieces, £3, 3s.

Phœbe.
By the Author of "Rutledge." Reprinted from the Fifth Thousand of the American Edition. Crown 8vo, 6s.

"'Phœbe' is a woman's novel."—*Saturday Review.*

Popular Genealogists;
Or, The Art of Pedigree-making. Crown 8vo, 4s.

The Gamekeeper's Manual: being Epitome of the Game
Laws for the use of Gamekeepers and others interested in the Preservation of Game. By Alexander Porter, Deputy Chief Constable of Roxburghshire. Fcap. 8vo, 1s.

Pictures from the Orkney Islands.
By John T. Reid, Author of "Art Rambles in Shetland." 4to, with numerous Illustrations, 25s.

Oils and Water Colours.
By William Renton. Fcap., 5s.

"The book is obviously for the Artist and the Poet, and for every one who shares with them a true love and zeal for nature's beauties."—*Scotsman.*

Kuram, Kabul, and Kandahar: being a Brief Record of
the Impressions in Three Campaigns under General Roberts. By Lieutenant Robertson, 8th, "The King's," Regiment. Crown 8vo, with Maps, 6s.

Scotland under her Early Kings.
A History of the Kingdom to the close of the 13th century. By E. William Robertson. In 2 vols. 8vo, cloth, 36s.

LIST OF BOOKS

Historical Essays,
In connection with the Land and the Church, etc. By E. WILLIAM ROBERTSON, Author of "Scotland under her Early Kings." 8vo, 10s. 6d.

A Rectorial Address delivered before the Students of Aberdeen University, in the Music Hall at Aberdeen, on Nov. 5, 1880. By THE EARL OF ROSEBERY. 6d.

A Rectorial Address delivered before the Students of the University of Edinburgh, Nov. 4, 1882. By THE EARL OF ROSEBERY. 6d.

Aberdour and Inchcolme. Being Historical Notices of the Parish and Monastery, in Twelve Lectures. By the Rev. WILLIAM ROSS, LL.D., Author of "Burgh Life in Dunfermline in the Olden Time." Crown 8vo, 6s.

Notes and Sketches from the Wild Coasts of Nipon.
With Chapters on Cruising after Pirates in Chinese Waters. By HENRY C. ST. JOHN, Captain R.N. Small Demy 8vo, with Maps and Illustrations, 12s.

"One of the most charming books of travel that has been published for some time."—*Scotsman.*

"There is a great deal more in the book than Natural History.... His pictures of life and manners are quaint and effective, and the more so from the writing being natural and free from effort."—*Athenæum.*

"He writes with a simplicity and directness, and not seldom with a degree of graphic power, which, even apart from the freshness of the matter, renders his book delightful reading. Nothing could be better of its kind than the description of the Inland Sea."—*Daily News.*

Notes on the Natural History of the Province of Moray.
By the late CHARLES ST. JOHN, Author of "Wild Sports in the Highlands." Second Edition. In 1 vol. Royal 8vo, with 40 page Illustrations of Scenery and Animal Life, engraved by A. DURAND after sketches made by GEORGE REID, R.S.A., and J. WYCLIFFE TAYLOR; also, 30 Pen-and-Ink Drawings by the Author in facsimile. 50s.

"This is a new edition of the work brought out by the friends of the late Mr. St. John in 1863; but it is so handsomely and nobly printed, and enriched with such charming illustrations, that we may consider it a new book."—*St. James's Gazette.*

"Charles St. John was not an artist, but he had the habit of roughly sketching animals in positions which interested him, and the present reprint is adorned by a great number of these, facsimiled from the author's original pen and ink. Some of these, as for instance the studies of the golden eagle swooping on its prey, and that of the otter swimming with a salmon in its mouth, are very interesting, and full of that charm that comes from the exact transcription of unusual observation."—*Pall Mall Gazette.*

A Tour in Sutherlandshire, with Extracts from the Field-Books of a Sportsman and Naturalist. By the late CHARLES ST. JOHN, Author of "Wild Sports and Natural History in the Highlands." Second Edition, with an Appendix on the Fauna of Sutherland, by J. A. HARVIE-BROWN and T. E. BUCKLEY. Illustrated with the original Wood Engravings, and additional Vignettes from the Author's sketch-books. In 2 vols. Small Demy 8vo, 21s.

"Every page is full of interest."—*The Field.*

"There is not a wild creature in the Highlands, from the great stag to the tiny fire-crested wren, of which he has not something pleasant to say."—*Pall Mall Gazette.*

Life of James Hepburn, Earl of Bothwell.
By Professor SCHIERN, Copenhagen. Translated from the Danish by the Rev. DAVID BERRY, F.S.A. Scot. Demy 8vo, 16s.

Scotch Folk.
Illustrated. Fourth Edition enlarged. Ex. fcap. 8vo, 1s.

"They are stories of the best type, quite equal in the main to the average of Dean Ramsay's well-known collection."—*Aberdeen Free Press.*

Studies in Poetry and Philosophy.

By J. C. SHAIRP, LL.D., Principal of the United College of St. Salvator and St. Leonard, St. Andrews. Third Edition. Crown 8vo,

"In the 'Moral Dynamic,' Mr. Shairp seeks for something which shall persuade us of the vital and close bearing on each other of moral thought and spiritual energy. It is this conviction which has animated Mr. Shairp in every page of the volume before us. It is because he appreciates so justly and forcibly the powers of philosophic doctrine over all the field of human life, that he leans with such strenuous trust upon those ideas which Wordsworth unsystematically, and Coleridge more systematically, made popular and fertile among us."—*Saturday Review.*

"The finest essay in the volume, partly because it is upon the greatest and most definite subject, is the first, on *Wordsworth*. . . . We have said so much upon this essay that we can only say of the other three that they are fully worthy to stand beside it."—*Spectator.*

Culture and Religion.

By PRINCIPAL SHAIRP. Seventh Edition. Fcap. 8vo, 3s. 6d.

"A wise book, and unlike a great many other wise books, has that carefully shaded thought and expression which fits Professor Shairp to speak for Culture no less than for Religion."—*Spectator.*

"Those who remember a former work of Principal Shairp's, 'Studies in Poetry and Philosophy,' will feel secure that all which comes from his pen will bear the marks of thought, at once careful, liberal, and accurate. Nor will they be disappointed in the present work. . . . We can recommend this book to our readers."—*Athenæum.*

"We cannot close without earnestly recommending the book to thoughtful young men. It combines the loftiest intellectual power with a simple and childlike faith in Christ, and exerts an influence which must be stimulating and healthful."—*Freeman.*

Kilmahoe, a Highland Pastoral,

And other Poems. By PRINCIPAL SHAIRP. Fcap. 8vo, 6s.

Shakespeare on Golf. With special Reference to St. Andrews Links. 3d.

The Divine Comedy of Dante Alighieri, The Inferno.

A Translation in Terza Rima, with Notes and Introductory Essay. By JAMES ROMANES SIBBALD. With an Engraving after Giotto's Portrait. Small Demy 8vo, 12s.

"Mr. Sibbald is certainly to be congratulated on having produced a translation which would probably give an English reader a better conception of the nature of the original poem, having regard both to its matter and its form in combination, than any other English translation yet published."—*Academy.*

The Use of what is called Evil.

A Discourse by SIMPLICIUS. Extracted from his Commentary on the Enchiridion of Epictetus. Crown 8vo, 1s.

The Near and the Far View,

And other Sermons. By Rev. A. L. SIMPSON, D.D., Derby. Ex. fcap. 8vo, 5s.

"Very fresh and thoughtful are these sermons."—*Literary World.*

"Dr. Simpson's sermons may fairly claim distinctive power. He looks at things with his own eyes, and often shows us what with ordinary vision we had failed to perceive. . . . The sermons are distinctively good."—*British Quarterly Review.*

Archæological Essays.

By the late Sir JAMES SIMPSON, Bart. Edited by the late JOHN STUART, LL.D. 2 vols. 4to, 21s.

1. Archæology.
2. Inchcolm.
3. The Cat Stane.
4. Magical Charm-Stones.
5. Pyramid of Gizeh.
6. Leprosy and Leper Hospitals.
7. Greek Medical Vases.
8. Was the Roman Army provided with Medical Officers?
9. Roman Medicine Stamps, etc. etc.

The Four Ancient Books of Wales,
Containing the Cymric Poems attributed to the Bards of the sixth century. By WILLIAM F. SKENE, D.C.L., Historiographer-Royal for Scotland. With Maps and Facsimiles. 2 vols. 8vo, 36s.

Celtic Scotland: A History of Ancient Alban.
By WILLIAM F. SKENE, D.C.L., Historiographer-Royal for Scotland. In 3 vols. Demy 8vo, 45s. Illustrated with Maps.
    I.—HISTORY and ETHNOLOGY.   II.—CHURCH and CULTURE.
               III.—LAND and PEOPLE.

"Forty years ago Mr. Skene published a small historical work on the Scottish Highlands which has ever since been appealed to as an authority, but which has long been out of print. The promise of this youthful effort is amply fulfilled in the three weighty volumes of his maturer years. As a work of historical research it ought in our opinion to take a very high rank."—*Times.*

The Gospel History for the Young:
Being lessons on the Life of Christ, Adapted for use in Families and Sunday Schools. By WILLIAM F. SKENE, D.C.L., Historiographer-Royal for Scotland. Small Crown 8vo, 3 vols., with Maps, 5s. each vol., or in cloth box, 15s.

"In a spirit altogether unsectarian, provides for the young a simple, interesting, and thoroughly charming history of our Lord."—*Literary World.*
"This 'Gospel History for the Young' is one of the most valuable books of the kind."—*The Churchman.*

Scottish Woodwork of the Sixteenth and Seventeenth
Centuries. Measured, Drawn, and Lithographed by J. W. SMALL, Architect. In one folio volume, with 130 Plates, Four Guineas.

Shelley: a Critical Biography.
By GEORGE BARNETT SMITH. Ex. fcap. 8vo, 6s.

The Sermon on the Mount.
By the Rev. WALTER C. SMITH, D.D. Crown 8vo, 6s.

Life and Work at the Great Pyramid.
With a Discussion of the Facts ascertained. By C. PIAZZI SMYTH, F.R.SS.L. and E., Astronomer-Royal for Scotland. 3 vols. Demy 8vo, 56s.

Madeira Meteorologic:
Being a Paper on the above subject read before the Royal Society, Edinburgh, on the 1st of May 1882. By C. PIAZZI SMYTH, Astronomer-Royal for Scotland. Small 4to, 6s.

Saskatchewan and the Rocky Mountains.
Diary and Narrative of Travel, Sport, and Adventure, during a Journey through part of the Hudson's Bay Company's Territories in 1859 and 1860. By the EARL OF SOUTHESK, K.T., F.R.G.S. 1 vol. Demy 8vo, with Illustrations on Wood by WHYMPER, 18s.

By the same Author.

Herminius:
A Romance. Fcap. 8vo, 6s.

Jonas Fisher:
A Poem in Brown and White. Cheap Edition. 1s.

The Burial of Isis and other Poems.
Fcap. 8vo, 6s.

Darroll, and other Poems.
By WALTER COOK SPENS, Advocate. Crown 8vo, 5s.

PUBLISHED BY DAVID DOUGLAS.

Rudder Grange.
By FRANK R. STOCKTON. 1 vol. 32mo, 1s.; and cloth, 2s.

"It may be safely recommended as a very amusing little book."—*Athenæum.*
"Altogether 'Rudder Grange' is as cheery, as humorous, and as wholesome a little story as we have read for many a day."—*St James's Gazette.*
"The minutest incidents are narrated with such genuine humour and gaiety, that at the close of the volume the reader is sorry to take leave of the merry innocent party."—*Westminster Review.*

The Lady or the Tiger? and other Stories.
By FRANK R. STOCKTON. 32mo, 1s.; and cloth, 2s.

Christianity Confirmed by Jewish and Heathen Testimony, and the Deductions from Physical Science, etc.
By THOMAS STEVENSON, F.R.S.E., F.G.S., Member of the Institution of Civil Engineers. Second Edition. Fcap. 8vo, 3s. 6d.

What is Play?
A Physiological Inquiry. Its bearing upon Education and Training. By JOHN STRACHAN, M.D. Fcap., 1s.

Good Lives: Some Fruits of the Nineteenth Century.
By A. M. SYMINGTON, D.D. Small Crown 8vo, 3s. 6d.

Sketch of Thermodynamics.
By P. G. TAIT, Professor of Natural Philosophy in the University of Edinburgh. Second Edition, revised and extended. Crown 8vo, 5s.

Talks with our Farm-Servants.
By An Old Farm-Servant. Crown 8vo; paper 6d., cloth 1s.

Walden; or, Life in the Woods.
By H. D. THOREAU. Crown 8vo, 6s.

Tommie Brown and the Queen of the Fairies; a new
Child's Book, in fcap. 8vo. With Illustrations, 4s. 6d.

Let pain be pleasure and pleasure be pain.

"There is no wonder that children liked the story. It is told neatly and well, and is full of great cleverness, while it has that peculiar character the absence of which from many like stories deprives them of any real interest for children."—*Scotsman.*

Our Mission to the Court of Marocco in 1880, under
Sir JOHN DRUMMOND HAY, K.C.B., Minister Plenipotentiary at Tangier, and Envoy Extraordinary to His Majesty the Sultan of Marocco. By Captain PHILIP DURHAM TROTTER, 93d Highlanders. Illustrated from Photographs by the Hon. D. LAWLESS, Rifle Brigade. Square Demy 8vo, 24s.

The Upland Tarn: A Village Idyll.
Small Crown, 5s.

Mr. Washington Adams in England.
By RICHARD GRANT WHITE. 32mo, 1s.; or in cloth, 2s.

"An impudent book."—*Vanity Fair.*
"This short, tiresome book."—*Saturday Review.*
"Brimful of genuine humour."—*Montrose Standard.*
"Mr. White is a capital caricaturist, but in portraying the ludicrous eccentricities of the patrician Britisher he hardly succeeds so well as in delineating the peculiar charms of the representative Yankee."—*Whitehall Review.*

The Botany of Three Historical Records:
Pharaoh's Dream, the Sower, and the King's Measure. By A. STEPHEN WILSON. Crown 8vo, with 5 plates, 3s. 6d.

LIST OF BOOKS PUBLISHED BY DAVID DOUGLAS.

"A Bushel of Corn."
By A. STEPHEN WILSON. An investigation by Experiments into all the more important questions which range themselves round a Bushel of Wheat, a Bushel of Barley, and a Bushel of Oats. Crown 8vo, with Illustrations, 9s.

"It is full of originality and force."—*Nature.*
"A monument of painstaking research."—*Liverpool Mercury.*
"Mr. Wilson's book is interesting not only for agriculturists and millers, but for all who desire information on the subject of corn, in which every one is so intimately concerned."—*Morning Post.*

Songs and Poems.
By A. STEPHEN WILSON. Crown 8vo, 6s.

Reminiscences of Old Edinburgh.
By DANIEL WILSON, LL.D., F.R.S.E., Professor of History and English Literature in University College, Toronto, Author of "Prehistoric Annals of Scotland," etc. etc. 2 vols Post 8vo, 15s.

The India Civil Service as a Career for Scotsmen.
By J. WILSON, M.A. 1s.

Christianity and Reason:
Their necessary connection. By R. S. WYLD, LL.D. Extra fcap. 8vo, 3s. 6d.

EDINBURGH: DAVID DOUGLAS, CASTLE STREET.

www.ingramcontent.com/pod-product-compliance
Lightning Source LLC
Chambersburg PA
CBHW032007230426
43672CB00010B/2279